T0259686

Springer Theses

Recognizing Outstanding Ph.D. Research

Aims and Scope

The series "Springer Theses" brings together a selection of the very best Ph.D. theses from around the world and across the physical sciences. Nominated and endorsed by two recognized specialists, each published volume has been selected for its scientific excellence and the high impact of its contents for the pertinent field of research. For greater accessibility to non-specialists, the published versions include an extended introduction, as well as a foreword by the student's supervisor explaining the special relevance of the work for the field. As a whole, the series will provide a valuable resource both for newcomers to the research fields described, and for other scientists seeking detailed background information on special questions. Finally, it provides an accredited documentation of the valuable contributions made by today's younger generation of scientists.

Theses are accepted into the series by invited nomination only and must fulfill all of the following criteria

- They must be written in good English.
- The topic should fall within the confines of Chemistry, Physics, Earth Sciences, Engineering and related interdisciplinary fields such as Materials, Nanoscience, Chemical Engineering, Complex Systems and Biophysics.
- The work reported in the thesis must represent a significant scientific advance.
- If the thesis includes previously published material, permission to reproduce this must be gained from the respective copyright holder.
- They must have been examined and passed during the 12 months prior to nomination.
- Each thesis should include a foreword by the supervisor outlining the significance of its content.
- The theses should have a clearly defined structure including an introduction accessible to scientists not expert in that particular field.

More information about this series at http://www.springer.com/series/8790

Takeshi Akuhara

Fluid Distribution Along the Nankai-Trough Megathrust Fault off the Kii Peninsula

Inferred from Receiver Function Analysis

Doctoral Thesis accepted by
The University of Tokyo, Tokyo, Japan

 Springer

Author
Dr. Takeshi Akuhara
Earthquake Research Institute
The University of Tokyo
Bunkyo-ku, Tokyo
Japan

Supervisor
Assoc. Prof. Kimihiro Mochizuki
Earthquake Research Institute
The University of Tokyo
Bunkyo-ku, Tokyo
Japan

ISSN 2190-5053   ISSN 2190-5061   (electronic)
Springer Theses
ISBN 978-981-13-4085-7   ISBN 978-981-10-8174-3   (eBook)
https://doi.org/10.1007/978-981-10-8174-3

Printed on acid-free paper

This Springer imprint is published by the registered company Springer Nature Singapore Pte Ltd.
part of Springer Nature
The registered company address is: 152 Beach Road, #21-01/04 Gateway East, Singapore 189721,
Singapore

Dedicated to my wife and family, who always support and encourage me.

Supervisor's Foreword

Dr. Akuhara first visited me during his search for potential major subjects for his graduate studies in the last year of his undergraduate studies. It was still six months before the 2011 Tohoku-Oki earthquake, the largest earthquake ever recorded in Japan. I remember asking him what subject attracted him most, and his answer was that he wanted to make an important contribution to mitigating disasters caused by large earthquakes (or megathrust events) along the Nankai Trough; these events had been drawing people's attention due to the high expectation of the occurrence such an earthquake in the near future. The Tohoku-Oki earthquake was not as important as those along the Nankai Trough.

Throughout history, we have repeatedly experienced large M 8 class earthquakes along the Nankai Trough running offshore of the southwestern region of Japan. Previous studies on the history of these large earthquakes suggest that their faults appear to be characterized by two major along-strike segments with their separating boundary being located off the Kii peninsula. The most recent events were the 1944 Tonankai earthquake to the east of the boundary, and the 1946 Nankai earthquake to the west. Considering a fairly constant recurrence interval of around 100 years, we should be well prepared for the next "megathrust" event.

I introduced him to the idea of a potential study of the structural factors that control the frictional behavior between subducting and overriding plates by analyzing a set of offshore seismic data collected by repeating observations using ocean-bottom seismometers (OBSs) around the major fault boundary of the repeating megathrusts along the Nankai Trough. Almost all of the megathrust events have occurred along the plate interface between the subducting and overriding plates beneath the offshore region. Therefore, offshore seismic data collected by OBSs are essential for understanding not only detailed seismicity but also subsurface structure around the megathrust faults. I also explained to him that although offshore seismic data is valuable, its quality cannot be compared with that of data from onshore seismic stations. The observation condition for OBSs on the seafloor is far from ideal: the instruments are placed over soft sediments, and the collapse of waves on the sea surface is actually a source of the noise—even for the quiet onshore stations. Therefore, a special process must be applied to the original OBS data before any analysis can be undertaken. However, instead of not wishing to face these difficulties, he rather found interests in tackling such problems by utilizing OBS data to allow for discussions of subsurface structure. In fact, his studies during his Ph.D. course are extremely relevant to the development of analysis methods utilizing OBS waveform records for the deduction of subsurface structures at depth in megathrust faults.

The aim of Dr. Akuhara's doctoral research was to reveal a seamless picture of structural changes along the subducting oceanic crust, and to propose a model of the subduction processes that includes variation in frictional properties with depth along the plate interface. He employed receiver function analysis applied to the OBS waveform records, which was considered a powerful tool for the deduction of changes in seismic properties across structural boundaries, such as the plate interface. The method has been widely applied to onshore seismic data. However, there had not yet been a successful application to OBS data due to interference from reverberations of seismic energy within the water column. Dr. Akuhara developed a method to filter out such reverberations so that quantitative discussion on seismic properties across structural interfaces beneath the seafloor was made possible. He combined results from receiver function analyses using onshore and offshore seismic data, and obtained a seamless picture of amplitude changes of the receiver function along the plate interface. He proposed a model that explains such amplitude changes by incorporating dehydration processes of the subducting oceanic crust and migration of the produced fluid, which then affect the frictional properties along the plate interface. Dr. Akuhara then further developed a method of receiver function analysis to deduce the km-scale structure around the plate interface at depth in megathrust faults, which has never been achieved by previous studies. He then succeeded in identifying a thin layer of low seismic velocities at the top of the subducting plate. From these results, he proposed a model of subduction processes.

His presentations of the results have been highly regarded, and he has been awarded Outstanding Student Paper Awards by the Japan Geoscience Union and the Seismological Society of Japan (SSJ). It is remarkable that both of his presentations at the 2015 SSJ meeting were awarded the Outstanding Student Paper Awards. Finally, his doctoral thesis was awarded the Research Encouragement Award by the Faculty of Science at the University of Tokyo. His results obtained by successful application of the methods developed through his doctoral study have been recognized by worldwide seismologists, and he has received offers of international collaborative research. He spent six months as a postdoctoral fellow at The University of British Colombia, Canada, on a research abroad program of the Japan Society for the Promotion of Science, where he started collaborative research on the structure of the Cascadia subduction zone. He is now an assistant professor at the Earthquake Research Institute at the University of Tokyo, and promotes the application of his original method to seismic data collected at worldwide subduction zones. It is expected that high-resolution pictures of plate interfaces will be obtained through the application of his method and a better understanding of the distribution of frictional properties along plate interfaces and the physical processes that determine such a distribution will be gained.

Tokyo, Japan   Assoc. Prof. Kimihiro Mochizuki
September 2017

Parts of this thesis have been published in the following journal articles:

- Akuhara T, Mochizuki K (2015) Hydrous state of the subducting Philippine Sea plate inferred from receiver function image using onshore and offshore data. *J Geophys Res Solid Earth* 120:8461–8477. doi: 10.1002/2015JB012336.
- Akuhara T, Mochizuki K, Kawakatsu H, Takeuchi N (2016) Non-linear waveform analysis for water-layer response and its application to high-frequency receiver function analysis using OBS array. *Geophys J Int* 206:1914–1920. doi: 10.1093/gji/ggw253.
- Akuhara T, Mochizuki K, Kawakatsu H, Takeuchi N (2017) A fluid-rich layer along the Nankai trough megathrust fault off the Kii Peninsula inferred from receiver function inversion. *J Geophys Res Solid Earth* 122:6524–6537. doi: 10. 1002/2017JB013965.

Acknowledgements

First and foremost, I would like to show my greatest appreciation to Assoc. Prof. Kimihiro Mochizuki who offered continuing support and constant encouragement throughout my Ph.D. course. This thesis would not have been possible without his help. My thanks also go to defense committee members of the corresponding Ph.D. dissertation: Profs. Satoshi Ide, Hitoshi Kawakatsu, Masanao Shinohara, and Assoc. Profs. Takashi Iidaka and Jin-Oh Park. Their comments helped me to improve the dissertation greatly. In addition, Prof. Hitoshi Kawakatsu gave me insightful suggestions on my research as a co-author of articles.

A lot of people helped me complete this book and I would like to express my gratitude to them: Dr. Takashi Tonegawa, for his giving me an introduction to receiver function analysis; Prof. Michael Bostock for discussion regarding low-velocity zones along subduction zones worldwide; Assoc. Prof. Nozomu Takeuchi for his assistance in developing a non-linear inversion code; Drs. Ayako Nakanishi, Yojiro Yamamoto, Katsuhiko Shiomi, and Prof. Takuo Shibutani for their discussion regarding subsurface structure around the Kii Peninsula; and colleagues at The University of Tokyo and The University of British Columbia for their sincere support.

The OBS data used in this study are highly admirable in terms of both their quality and quantity, and were collected by Profs. Masanao Shinohara, Hiroshi Shimizu, Assoc. Profs. Kimihiro Mochizuki, Shin'ichi Sakai and Kazuo Nakahigashi, and Drs. Tomoaki Yamada, Toshihiko Kanazawa, and Kenji Uehira. I would like to express my gratitude to them.

I also wish to thank the Japan Meteorology Agency, National Research Institute for Earth Science and Disaster Prevention, The University of Tokyo, Kochi University, and Kyoto University for providing continuous waveform data. We used The Generic Mapping Tools for drawing figures. This thesis also makes use of the computer package NA which was made available with support from the Inversion Laboratory (ilab). Ilab is a program for the construction and distribution of data inference software in the geosciences supported by AuScope Ltd, a non-profit organization for Earth Science infrastructure funded by the Australian Federal Government. This work was supported by JSPS KAKENHI Grant Number 26-10221.

Last but not least, I would also like to express my sincere gratitude to my wife and family for their moral support and warm encouragement.

Contents

Chapter 1
General Introduction

Abstract The study of fluid distribution along megathrust faults at subduction zones is important because such fluid controls the slip behavior of the megathrust faults. A number of studies have directed their efforts toward understanding the fluid distribution, exploiting both active- and passive-source seismic surveys. Receiver function (RF) analysis is a powerful method for this purpose. The method aims to extract P-to-S conversion phases from seismograms, which are useful for constraining physical properties along subsurface discontinuities such as subducting plate interfaces. While megathrust faults are located beneath offshore regions, RF analysis has hardly been applied to data recorded by ocean-bottom seismometers (OBSs) because seismograms recorded at offshore observatories are dominated by water reverberations. This thesis develops a new technique for solving this problem and then uses the technique to reveal the subsurface structure and fluid distribution along the megathrust fault off the Kii Peninsula, southwestern Japan. This chapter provides the background to this study, the tectonic setting of the study area, and descriptions of the data used in this thesis.

Keywords Subduction zones · Megathrust faults · Fluid distribution
Receiver function · Ocean-bottom seismometer

1.1 Fluid Distribution Along Megathrust Faults

Subduction zones host megathrust earthquakes that occasionally result in tens of thousands of fatalities caused by strong ground motion and tsunamis. Megathrust earthquakes have a wide rupture area on megathrust faults but the spatial extent in the along-dip direction is limited to the region of high seismic coupling (Fig. 1.1). Such an area is often defined as the seismogenic zone, typically ranging from 10 to 40 km depth, although its updip limit is not well constrained. Aseismic slip occurs on the plate interface at greater depths with almost no mechanical coupling between the subducting and overriding plates. The middle region between the seismogenic and aseismic zones is known as the transition zone, where various types of slow

© Springer Nature Singapore Pte Ltd. 2018   1
T. Akuhara, *Fluid Distribution Along the Nankai-Trough Megathrust Fault off the Kii Peninsula*, Springer Theses, https://doi.org/10.1007/978-981-10-8174-3_1

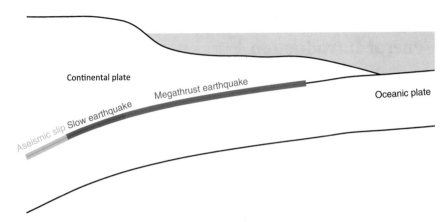

Fig. 1.1 Schematic illustration of a subduction zone. Along the plate interface, different types of fault slip occur in accordance with the interface depths

earthquakes, including non-volcanic tremors, low-frequency earthquakes (LFEs), and slow slip events (SSEs), occur.

Subduction zones play a role in the transportation of water into the Earth's deep interior. The subducting oceanic plate bends downward right before the onset of its subduction to form bend-related faults, and seawater seeps into the oceanic plate along these faults (Faccenda et al. 2009). Fluid incorporated in the oceanic plate in this way is released from the subducting plates during subduction via mechanical compaction and metamorphic dehydration reactions (Peacock and Wang 1999; Saffer and Tobin 2011). It is considered that some of the fluid is trapped along the plate interface, reducing fault strength (Scholz 1998) and causing non-volcanic tremors (Fagereng and Diener 2011; Shelly et al. 2006) and SSEs (Kato et al. 2010; Kodaira et al. 2004; Song et al. 2009). Moreover, such fluid might control even the rupture zone of megathrust earthquakes (Kimura et al. 2012; Zhao et al. 2011). However, the detailed fluid distribution along megathrust faults and its relationship to the slip behavior of such faults remain unclear.

To date, a number of seismological studies have directed their efforts toward revealing the physical properties of subducting plates using both active and passive seismic sources. These studies have detected low-velocity zones (LVZs) and/or high-reflectivity zones along the subducting plate boundaries of many subduction zones around the world (Table 1.1). Both LVZs and high-reflectivity zones are often considered as evidence of abundant fluid along the plate interface. This is because fluid reduces seismic velocities compared with surrounding rocks and yields a strong impedance contrast leading to intense reflection phases. One of the prominent contributions to this interpretation was by Audet et al. (2009), who investigated LVZ properties beneath the northern Cascadia subduction zone using a so-called receiver function (RF) technique. Their estimation of an anomalously high Poisson's ratio (~0.4) for the LVZ strongly suggests that the LVZ reflects overpressured oceanic crust. In addition, waveform modeling studies of reflection phases have revealed that high-reflectivity zones have low velocity, suggesting fluid-rich conditions (Bangs et al. 2009; Kodaira et al. 2002; Li et al. 2015).

As tools for investigating fluid distribution, active- and passive-source seismic explorations have both advantages and disadvantages. Active-source surveys have usually been performed along straight survey lines with densely deployed seismic stations. They have elucidated P-wave (i.e., compressional wave, also known as primary wave) velocity structure at shallow depths, or in seismogenic zones, using P-to-P reflection and/or P refraction phases. The high-frequency content in source wavelets and close separation distance of receivers enables the production of fine-scale (~100 m) images of subsurface structures. Trapped fluid around the plate interface has been reported based on intense P-to-P reflection phases (e.g., Bangs et al. 2009; Bell et al. 2010; Kodaira et al. 2002; Li et al. 2015; Mochizuki et al. 2005; Nedimović et al. 2003). The drawbacks of the active-source surveys are insensitivity to S-wave velocity (i.e., shear wave, also known as secondary wave), poor resolution of deep structures, and difficulty in performing 3-D analysis. In particular, knowledge of S-wave velocity, or Poisson's ratio, is essential to assessing rock types and fluid content (Christensen 1984, 1996).

On the other hand, passive-source surveys have revealed the hydrous state of subducting plates at deeper depths, mostly deeper than the seismogenic zone, through tomographic or receiver function (RF) analyses. A number of tomographic analyses have identified LVZs (for both P- and S-wave velocities) with high Poisson's ratios along subducting plates, which are often interpreted as hydrated oceanic crust (e.g., Hirose et al. 2008; Husen and Kissling 2001; Reyners and Eberhart-Phillips 2009; Tsuji et al. 2008). In general, however, the spatial resolution of these tomographic analyses is much lower than that of active-source surveys because of the sparse distribution of seismic stations. Moreover, smoothing constraints incorporated in tomographic analysis cause underestimation of the magnitudes of velocity anomalies (Song and Helmberger 2007). These drawbacks make it somewhat difficult to evaluate the correct thickness and velocity of the LVZ.

RF studies have also identified LVZs by detecting P-to-S conversion phases at both sides of a LVZ (e.g., Abe et al. 2013; Audet et al. 2009; Hansen et al. 2012; Kawakatsu and Watada 2007; Kim et al. 2010). In principle, RF analysis has the potential to achieve finer spatial and vertical resolutions than tomographic analysis, although many RF analyses may fail to estimate correct LVZ thicknesses due to simplistic velocity models used in time-to-depth conversion of RFs, as noted by by Bostock (2013). The RF amplitudes of P-to-S conversion phases provide information about S-wave velocity. In addition, P-wave velocity and density can be constrained by employing reflection phases as analysis targets. In spite of these promising features, RF analysis has thus far been rarely applied to shallow subduction zones, mainly because of the lack of offshore observations and difficulties in analyzing offshore data.

As we have seen above, previous active- and passive-source surveys exhibit differences in terms of the main target depth, sensitivity, and spatial resolution. These differences prohibit making a unified or seamless interpretation about fluid distribution from a seismogenic zone to a transition zone. One way to overcome this problem

Table 1.1 Low-velocity and high-reflectivity zones along the subducting plate interfaces worldwide

Region	Thickness (km)	Vp (km/s)	Vs(km/s)	Depth (km)	Key observation	Reference
SW Japan	–	3	–	15–30	Reflection phase from active source	Kodaira et al. 2002
	–	–	–	20–35	Reflection phase from active source	Kurashimo et al. 2013
	1–2	–	–	<8	Reflection phase from active source	Bangs et al. 2009
NE Japan	0.1–0.4	<4	–	10–20	Reflection phase from active source	Mochizuki et al. 2005
	–	–	~ −10%	40–90	Receiver function	Kawakatsu and Watada 2007
Cascadia	2–4	2.0–3.5 (Vp/Vs)		20–40	Receiver function	Hansen et al. 2012
	<2	–	–	15–20	Reflection phase from active source	Nedimović et al. 2003
	4>	–	–	20–40	Reflection phase from active source	
Mexico	3–5	–	2.0–2.7	20–45	Reflection phase from passive source	Song et al. 2009
Alaska	0.10–0.2	–	–	13–20	Reflection phase from active source	Li et al. 2015
	3–5	–	–	25–55	Reflection phase from active source	
Nazca	–	–	2.5–4.0	50–150	Receiver function	Kim and Clayton 2015
Southern ecuador	1	2.7	–	5–7	Reflection nad refraction phases of active source	Calahorrano et al. 2008
Costa Rica	3–5	1.9–2.9 (Vp/Vs)		15–30	Receiver function	Audet and Schwartz 2013
SW Japan (this study)	–	–	2.9–4.2	20–35	Receiver function	This study (Chap. 4)
	0.2–1.2	–	0.7–2.4	15–20	Receiver function	This study (Chap. 5)

Fig. 1.2 Teleseismic waveforms recorded by vertical-component sensors of OBSs. Waveforms are aligned with their largest amplitudes (dashed line). Green triangles indicate the expected timings of water reverberations, which were calculated using twice of the OBS depth divided by the P-wave velocity in the water layer (1.5 km/s)

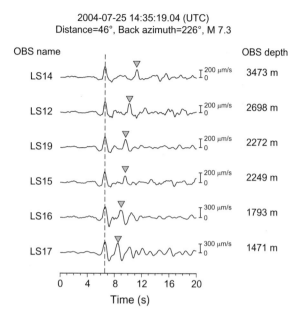

is to perform RF analysis using OBS data. In this thesis, we first aim to develop a RF method applicable to OBSs and demonstrate the potential of RF analysis with even complex OBS data. The other aim of this thesis is to investigate fluid distribution along the Nankai Trough megathrust fault off the Kii Peninsula, southwestern Japan, and to discuss its relationship with slip behavior on the megathrust fault.

A problem specific to OBS records is presence of the seawater. In general, a RF is estimated by deconvolving a horizontal-component record with its source wavelet, which is usually approximated by its vertical-component record. This approximation, however, is unsatisfactory for OBS data because strong water reverberations appear on the vertical-component records (Fig. 1.2). In Chap. 2, we introduce a water-layer filter (WLF) method, in which we take a series of reverberations in the water column as a linear filter. Using its inverse filter, we can then eliminate water reverberations. In Chap. 3, we apply this method to observed data to demonstrate the efficiency of our method and introduce a method for estimating the parameters required by the WLF.

In Chap. 4, we investigate the subduction zone structure and its hydrous state beneath the onshore and offshore regions around the Kii Peninsula, southwestern Japan. We perform a common-convergence-point stacking method to create RF image of subsurface structure. From the image, we first estimate the 3-D geometry of the subducting Philippine Sea (PHS) Plate and then estimate fluid content along the subducting plate. Although this method cannot accurately evaluate the thickness of the LVZ, we can still obtain useful information from the RF amplitudes to acquire an intuitive understanding of fluid distribution along the subducting plate. We make

an interpretation of the dehydration process of the subducting oceanic crust based
on RF amplitudes, seismicity, and a previous tomography model.

Chapter 5 aims to quantitatively assess the detailed features of the LVZ beneath
the offshore region around the Kii Peninsula. For this purpose, we calculate RFs using
a higher-frequency range (<4.0 Hz). We first evaluate the properties of the sediment
layer beneath the seafloor by a simple stacking method and then investigate the deeper
structure by performing RF inversion analysis. Our results show that the LVZ forms
an extremely thin layer, which reflects a fluid-rich sediment layer along the plate
interface.

In Chap. 6, we briefly provide general discussion derived from this thesis. First,
we roughly estimate the spatial variation in LVZ thickness and speculate as to the
dehydration process and fluid distribution along the subducting plate from the seismogenic zone to the transition zone. Then, we compare what we have learned from
our study with that learned from other subduction zones and discuss the design of
future subduction zone studies aimed at better understand the relationship between
fluid distribution and slip behavior on megathrust faults.

1.2   Receiver Function Analysis

When a teleseismic P-wave encounters a seismic velocity discontinuity (plate interfaces, Moho, etc.), part of the energy is converted into an S-wave. Since the propagation velocities of a P-wave are higher than those of an S-wave, P-to-S conversion
phases arrive later than the direct P phase. Provided the seismic velocity is known,
measuring the differential arrival time between the P-to-S and P phase enables estimation of the depths of the velocity discontinuity where P-to-S conversion occurs.
In addition, the efficiency of the conversion depends on S-wave velocity contrast:
more efficient conversion occurs at stronger velocity contrast. Hence, the relative
amplitudes of P-to-S conversion to the P phase can be used to constrain S-wave
velocity.

However, the onsets of P-to-S phases are not generally visible on actual seismograms because they are overprinted by long-tailed coda of the direct P arrival
(Fig. 1.3c). RF analysis, which will be introduced below, is widely used in the field
of seismology to overcome this difficulty.

Three-component seismograms of teleseismic waveforms are expressed as:

$$u_v(t) = s(t) * G_v(t) \tag{1.1}$$
$$u_r(t) = s(t) * G_r(t) \tag{1.2}$$
$$u_t(t) = s(t) * G_t(t), \tag{1.3}$$

where $u_v(t)$, $u_r(t)$, and $u_t(t)$ represent vertical-, radial-, and transverse-component
records (Fig. 1.3c), respectively, $s(t)$ represents a source wavelet incidenting from
below a receiver-side structure (Fig. 1.3a), and $G_v(t)$, $G_r(t)$, and $G_t(t)$ represent

Fig. 1.3 Schematic illustration of seismograms. Teleseismic waveform records can be expressed as the convolution of an incident source wavelet and Green's function (i.e., an impulse response) of a receiver-side structure. Each pulse appearing on a Green's function corresponds to a phase arrival, such as a direct P arrival or P-to-S conversions from subsurface interfaces

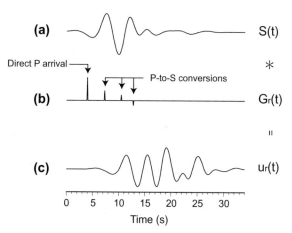

vertical-, radial-, and transverse-component Green's functions (i.e., impulse responses) of the receiver-side structure, respectively (Fig. 1.3b). Asterisks (∗) in Eqs. 1.1–1.3 represent convolution operations. Note that in Eqs. 1.1–1.3, instrumental responses and noise terms are neglected for simplicity. Our goal is to retrieve $G_r(t)$ because P-to-S conversions have the strongest amplitude on the radial component. For this purpose, the source wavelet term is often approximated by the vertical-component record, i.e., $S(t) \simeq G_v(t)$. This approximation enables calculation of $G_r(t)$ by deconvolution of a radial-component record by the vertical-component record.

$$G_r(t) = u_r(t) * \{u_v(t)\}^{-1} \qquad (1.4)$$

The resultant $G_r(t)$ is termed the radial-component RF (the transverse-component RF can be calculated from $u_t(t)$ in the same manner).

1.3 Tectonic Setting of the Nankai Subduction Zone

Our study area extends from the onshore to offshore regions around the Kii Peninsula, located in the central part of the southwestern (SW) Japan subduction zone (Fig. 1.4). Here the PHS plate subducts along the N55W° direction at a convergence rate of 63–68 mm/yr relative to the Amurian plate (Miyazaki and Heki 2001). The onset of the subduction is located ∼100 km off the Kii Peninsula at the Nankai Trough. The PHS plate was formed by the back-arc opening of the Sikoku Basin between 27 and 15 Ma (Okino et al. 1999). For its young age, the temperature of the PHS plate is relatively high in comparison with other subduction zones around the world. Along the fossil spreading axis, the Kinan Seamount Chain is located; its northern end has been already subducted.

8   1   General Introduction

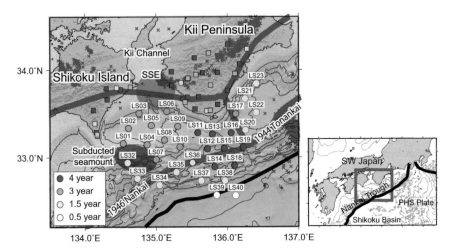

Fig. 1.4 Tectonic setting and station distribution of our study area. Circles denote ocean-bottom seismometers (color denotes observational period). Squares are on-land permanent stations operated by the National Research Institute for Earth Science and Disaster Prevention (red), Japan Meteorological Agency (yellow), University of Tokyo (light green), Kyoto University (blue), and Kochi University (sky-blue). The thick black curve represents the Nankai Trough. Brown- and pink-shaded areas represent a subducted seamount (Kodaira et al. 2002) and major slip zone of the 1996–1997 slow slip event (Kobayashi 2014), respectively. Blue crosses show non-volcanic tremors that occurred from January 1, 2004 to December 31, 2004 (Idehara et al. 2014). Thick blue curves roughly enclose the rupture area of the 1944 Tonankai and 1946 Nankai earthquakes (Baba et al. 2006, 2002). The lower-right insert indicates the regional tectonic setting. Modified from Akuhara and Mochizuki (2015) by permission of John Wiley & Sons Ltd

Megathrust earthquakes have occurred on the plate interface between the subuducting PHS plate and the overriding plate on 100- to 150-year cycle (Ando 1975). The two most recent events were the 1944 Tonankai and 1946 Nankai earthquakes. A number of studies have shown that the rupture area boundary of these two events was located at the southern tip of the Kii Peninsula (e.g., Baba et al. 2006, 2002). Historical records suggest that some of previous events had the same rupture area boundary (Ando 1975), so there should be some structural factors. A number of candidates have been suggested as the cause of the rupture area boundary, such as changes in the thickness of the oceanic crust (Mochizuki et al. 1998), the vertical load caused by a dense rock body embedded in the overriding plate (Kodaira et al. 2006), and lateral variation in pore fluid pressure (Kurashimo et al. 2013). A subducted seamount (brown area in Fig. 1.4) has also been considered to affect rupture propagation (Kodaira et al. 2000). However, the rupture processes of the past megathrust earthquakes are not fully understood, partly due to an insufficient number of seismic observatories at the time.

The downdip limit of the rupture zone is considered to be ∼25 km depth in this region. At greater downdip (i.e., the transition zone), non-volcanic tremors occur on the plate interface (Fig. 1.4, blue crosses). The source region of the tremors is limited in the along-dip direction, forming a belt-like zone of epicenters. The cause of this distribution is still controversial: One possibility is that the zone is under favorable pressure–temperature conditions for dehydration reactions of the oceanic crust to occur (Fagereng and Diener 2011), and another possibility is that the island-arc Moho intersecting the plate interface at the source region of tremors acts as an impermeable barrier to elevate pore fluid pressure (Katayama et al. 2012). The tremors are considered to be collections of LFEs (Shelly et al. 2007) and are accompanied by short-term SSEs, which rupture at the same location (Obara et al. 2004). A large gap in the tremor belt exists in the Kii Channel, where a recent study has revealed that a long-term SSE occurred from 1996 to 1997 (Kobayashi 2014). The reason a long-term SSE, rather than tremors, occured is not yet fully understood.

Passive seismic monitoring with a large number of OBSs was conducted from November 2004 to November 2007 in this region (circles in Fig. 1.4). Mochizuki et al. (2010) and Akuhara et al. (2013) revealed along-strike variation of seismicity and seismic velocity structure, respectively, using these OBS data. They inferred that such heterogeneities may be related to the opening history of the Shikoku Basin. Similar along-strike variation has been reported across onshore and offshore regions involved with various kinds of geophysical observations (Ide et al. 2010; Ishise et al. 2009; Park et al. 2014; Shiomi and Park 2008). Although the cause of the along-strike variation is still open to question, these observations highlight the necessity for 3-D analysis for this region.

A number of studies have focused on fluid content along the plate interface in this region, especially for shallow depths near the trench. Borehole experiments operated by Ocean Drilling Program have provided quantitative constraints on properties of décollement (e.g., permeability and pore fluid pressure). For example, porosity profiles of the boreholes have suggested overpressured décollement at the sites (Screaton et al. 2002). Furthermore, results of numerical modeling have reported that such an overpressured décollement zone extends from the trench to 40 km landward, corresponding to a depth of $\sim$5 km beneath the seafloor (Skarbek and Saffer 2009). The excess pore fluid pressure is considered to be caused by clay dehydration (Saffer and Tobin 2011). An issue in these studies based on borehole measurements lies in the sparseness of sampling sites, which hampers a detailed investigation of lateral heterogeneities. Park et al. (2014) investigated along-trench variation of décollement reflectivity by conducting active-source surveys and inferred that the décollement is segmented along the trench-parallel direction by fluid-rich and fluid-poor zones.

In contrast to the shallow region, fluid distribution at the seismogenic zone depth of this region ($\sim$10–25 km depth) has not been investigated well. One of the few studies was by Kodaira et al. (2002) who reported the presence of a fluid-rich sediment layer along the plate interface off the Shikoku Island based on intense P-to-P reflection phases. Since their survey line was aligned in the trench-normal direction, the lateral extent of the fluid-rich layer remains unclear, elucidating this is a goal of the present study.

On the Kii Peninsula, active- and passive-source surveys with densely aligned seismic stations have been conducted. Kurashimo et al. (2013) identified a high reflectivity band and interpreted it as excess pore fluid pressure along the plate interface.

Based on the results of tomography analysis, Kato et al. (2014) also inferred that the top of the subducting plate is hydrated. Both of these studies suggested that the fluid originates from the dehydration reaction of the subducting plate.

1.4 Data

1.4.1 Seismic Stations

In this study, we used waveform data obtained by OBSs deployed between Shikoku Island and the Kii Peninsula from November 2003 to November 2007 (circles in Fig. 1.4). These OBSs were equipped with three-component velocity sensors with a natural frequency of 1 Hz (Lennartz LE-3Dlite seismometers) and they recorded continuous waveform data. Because their battery capacity was limited to about one year, short-term observations spanning several months to a year were repeated. In total, OBSs were deployed at 32 sites on the seafloor, although the durations of the observational records varied among the sites (Mochizuki et al. 2010). The close proximity of the OBSs, the long observational periods, and the large number of observational sites have made this study area one of the best places in the world to demonstrate the potential of RF analysis with OBS data.

For installation, OBSs were released close to the sea surface and allowed to sink freely to the seafloor. The direction of the vertical-component sensor was adjusted automatically via a gimbal system, while the direction of the two horizontal component sensors remained unknown. We determined the orientations of the horizontal component sensors based on the polarities of Rayleigh waves, using a method similar to that of Baker and Stevens (2004). The averaged and maximum 1σ standard errors of the estimated orientations among all OBSs were 2° and 6°, respectively, which is reasonably small for RF analysis. We determined the sensor orientations for each OBS reinstallation. Technically, the OBS positions (latitude, longitude, and depth) also differed (in order of several hundred meters) after the reinstallation. However, we ignored this difference and instead used averaged positions. We confirmed that RFs calculated using earthquakes located close to each other showed no significant difference even if the OBSs were replaced.

In addition to the OBS data, we also used continuous waveform data obtained at on-land permanent seismic stations operated by the National Research Institute for Earth Science and Disaster Prevention (Hi-net stations), the Japan Meteorological Agency (JMA), and several universities (squares in Fig. 1.4). Although the instrumental responses of these stations differed slightly, their natural frequencies were 1.0 Hz, the same as those of OBSs.

1.4.2 Teleseismic Event Records

We extracted event records from the continuous waveform data based on two earth-quake catalogs: the International Seismological Centre-Global Earthquake Model (ISC-GEM) catalog and the JMA catalog. Referring to the ISC-GEM catalog, we selected 377 teleseismic events of $M_W \geq 6.0$ that occurred at distances of 30–90° from our study area between November 1, 2003 and December 31, 2007. From the JMA catalog, we chose 209 deep local events of Mj 5.0 that occurred at a depth of > 300 km depth within distances range 3–10° between November 1, 2003 and December 31, 2007. Based on the IASP91 Earth structure model (Kennett and Eng-dahl 1991), we calculated a synthetic P-wave travel time for each selected event to extract event records. We then rotated these event records to align them with the vertical, radial, and transverse coordinates, and resampled them at a 20-Hz sampling rate after the application of a 9.0-Hz low-pass filter to avoid aliasing.

Among these preprocessed event records, we selected those with a signal-to-noise ratio (SNR) of >3.0 on vertical components. The SNR was defined by the root-mean-square amplitude ratio calculated with 30-s time windows before and after the synthetic arrival times. Hereafter, we refer to this criterion as a "prerequisite crite-rion". Throughout this thesis, we further impose additional criteria as necessary. The number of event records satisfying the prerequisite criteria differs among stations. Roughly speaking, the number at on-land stations is 5–10 times larger than that at OBSs (Fig. 1.5).

Figure 1.6 shows the frequency-dependent SNR of the event records satisfying the prerequisite criteria. To produce this figure, we first calculated the Fourier amplitude

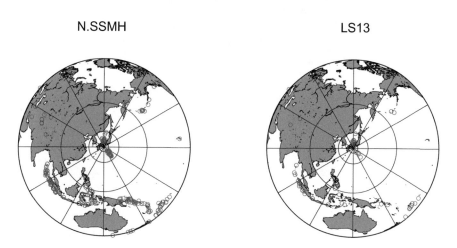

Fig. 1.5 Locations of teleseismic and deep events satisfying prerequisite criteria (see Sect. 1.4.2) for a typical on land station (left) and an ocean-bottom seismometer (right). Gray circles show 3°, 10°, and 30° distance from our study area

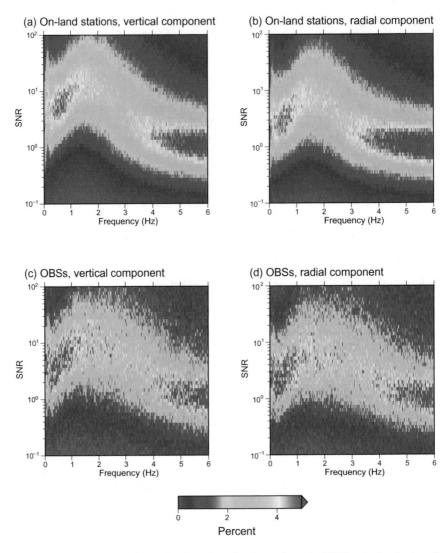

Fig. 1.6 Histogram of the frequency-dependent signal-to-noise ratio (SNR) calculated using all event records from (**a**, **b**) on-land stations and (**c**, **d**) OBSs of (**a**, **c**) vertical and (**b**, **d**) radial components. All event records used here satisfy the prerequisite criteria introduced in Sect. 1.4.2

spectrum of all event records satisfying the prerequisite criteria using the 30-s time windows before and after the theoretical arrival time. We then calculated the spectrum ratio of the two time windows to calculate the SNR for each frequency. In response to the magnitudes, the resultant frequency-dependent SNRs were divided into 50 bins with even intervals on a logarithmic scales. Finally, we counted the number of event records included in these bins and present it as a percentage (Fig. 1.6). We can observe that high SNRs are produced between ~0.5 Hz and ~3.5 Hz for both on-land stations and OBSs.

References

Abe Y, Ohkura T, Hirahara K, Shibutani T (2013) Along-arc variation in water distribution in the uppermost mantle beneath Kyushu, Japan, as derived from receiver function analyses. J Geophys Res Solid Earth 118(7):3540–3556. https://doi.org/10.1002/jgrb.50257

Akuhara T, Mochizuki K (2015) Hydrous state of the subducting philippine sea plate inferred from receiver function image using onshore and offshore data. J Geophys Res Solid Earth 120(12):8461–8477. https://doi.org/10.1002/2015JB012336

Akuhara T, Mochizuki K, Nakahigashi K, Yamada T, Shinohara M, Sakai S, Kanazawa T, Uehira K, Shimizu H (2013) Segmentation of the Vp/Vs ratio and low-frequency earthquake distribution around the fault boundary of the Tonankai and Nankai earthquakes. Geophys Res Lett 40(7):1306–1310. https://doi.org/10.1002/grl.50223

Ando M (1975) Source mechanisms and tectonic significance of historical earthquakes along the Nankai Trough, Japan. Tectonophysics 27(2):119–140. https://doi.org/10.1016/0040-1951(75)90102-X

Audet P, Schwartz SY (2013) Hydrologic control of forearc strength and seismicity in the Costa Rican subduction zone. Nat Geosci 6(10):852–855. https://doi.org/10.1038/ngeo1927

Audet P, Bostock MG, Christensen NI, Peacock SM (2009) Seismic evidence for overpressured subducted oceanic crust and megathrust fault sealing. Nature 457(7225):76–78. https://doi.org/10.1038/nature07650

Baba T, Tanioka Y, Cummins PR, Uhira K (2002) The slip distribution of the 1946 Nankai earthquake estimated from tsunami inversion using a new plate model. Phys Earth Planet Inter 132(1–3):59–73. https://doi.org/10.1016/S0031-9201(02)00044-4

Baba T, Cummins PR, Hori T, Kaneda Y (2006) High precision slip distribution of the 1944 Tonankai earthquake inferred from tsunami waveforms: possible slip on a splay fault. Tectonophysics 426(1–2):119–134. https://doi.org/10.1016/j.tecto.2006.02.015

Baker GE, Stevens JL (2004) Backazimuth estimation reliability using surface wave polarization. Geophys Res Lett 31(9):L09,611. https://doi.org/10.1029/2004GL019510

Bangs N, Moore G, Gulick S, Pangborn E, Tobin H, Kuramoto S, Taira A (2009) Broad, weak regions of the Nankai Megathrust and implications for shallow coseismic slip. Earth Planet Sci Lett 284(1–2):44–49. https://doi.org/10.1016/j.epsl.2009.04.026

Bell R, Sutherland R, Barker DHN, Henrys S, Bannister S, Wallace L, Beavan J (2010) Seismic reflection character of the Hikurangi subduction interface, New Zealand, in the region of repeated Gisborne slow slip events. Geophys J Int 180(1):34–48. https://doi.org/10.1111/j.1365-246X.2009.04401.x

Bostock M (2013) The Moho in subduction zones. Tectonophysics 609:547–557. https://doi.org/10.1016/j.tecto.2012.07.007

Calahorrano A, Sallarès V, Collot JY, Sage F, Ranero CR (2008) Nonlinear variations of the physical properties along the southern Ecuador subduction channel: results from depth-migrated seismic data. Earth Planet Sci Lett 267(3–4):453–467. https://doi.org/10.1016/j.epsl.2007.11.061

Christensen NI (1984) Pore pressure and oceanic crustal seismic structure. Geophys J Int 79(2):411–423. https://doi.org/10.1111/j.1365-246X.1984.tb02232.x

Christensen NI (1996) Poisson's ratio and crustal seismology. J Geophys Res 101(B2):3139–3156. https://doi.org/10.1029/95JB03446

Faccenda M, Gerya TV, Burlini L (2009) Deep slab hydration induced by bending-related variations in tectonic pressure. Nat Geosci 2(11):790–793. https://doi.org/10.1038/ngeo656

Fagereng Å, Diener JFA (2011) Non-volcanic tremor and discontinuous slab dehydration. Geophys Res Lett 38(15):L15,302. https://doi.org/10.1029/2011GL048214

Hansen RT, Bostock MG, Christensen NI (2012) Nature of the low velocity zone in Cascadia from receiver function waveform inversion. Earth Planet Sci Lett 337–338:25–38. https://doi.org/10.1016/j.epsl.2012.05.031

Hirose F, Nakajima J, Hasegawa A (2008) Three-dimensional seismic velocity structure and configuration of the Philippine Sea slab in southwestern Japan estimated by double-difference tomography. J Geophys Res 113(B9):B09,315. https://doi.org/10.1029/2007JB005274

Husen S, Kissling E (2001) Postseismic fluid flow after the large subduction earthquake of Antofagasta, Chile. Geology 29(9):847–850. https://doi.org/10.1130/0091-7613(2001)029<0847:PFFATL>2.0.CO;2

Ide S, Shiomi K, Mochizuki K, Tonegawa T, Kimura G (2010) Split Philippine Sea plate beneath Japan. Geophys Res Lett 37(21):L21,304. https://doi.org/10.1029/2010GL044585

Idehara K, Yabe S, Ide S (2014) Regional and global variations in the temporal clustering of tectonic tremor activity. Earth Planets Space 66(1):66. https://doi.org/10.1186/1880-5981-66-66

Ishise M, Koketsu K, Miyake H (2009) Slab segmentation revealed by anisotropic P wave tomography. Geophys Res Lett 36. https://doi.org/10.1029/2009GL037749

Katayama I, Terada T, Okazaki K, Tanikawa W (2012) Episodic tremor and slow slip potentially linked to permeability contrasts at the Moho. Nat Geosci 5(10):731–734. https://doi.org/10.1038/ngeo1559

Kato A, Iidaka T, Ikuta R, Yoshida Y, Katsumata K, Iwasaki T, Sakai S, Thurber C, Tsumura N, Yamaoka K, Watanabe T, Kunitomo T, Yamazaki F, Okubo M, Suzuki S, Hirata N (2010) Variations of fluid pressure within the subducting oceanic crust and slow earthquakes. Geophys Res Lett 37(14):L14,310. https://doi.org/10.1029/2010GL043723

Kato A, Saiga A, Takeda T, Iwasaki T, Matsuzawa T (2014) Non-volcanic seismic swarm and fluid transportation driven by subduction of the Philippine Sea slab beneath the Kii Peninsula, Japan. Earth Planets Space 66(1):86. https://doi.org/10.1186/1880-5981-66-86

Kawakatsu H, Watada S (2007) Seismic evidence for deep-water transportation in the mantle. Science 316:1468–1471. https://doi.org/10.1126/science.1140855

Kennett BLN, Engdahl ER (1991) Traveltimes for global earthquake location and phase identification. Geophys J Int 105(2):429–465. https://doi.org/10.1111/j.1365-246X.1991.tb06724.x

Kim Y, Clayton RW (2015) Seismic properties of the Nazca oceanic crust in southern Peruvian subduction system. Earth Planet Sci Lett 429:110–121. https://doi.org/10.1016/j.epsl.2015.07.055

Kim Y, Clayton RW, Jackson JM (2010) Geometry and seismic properties of the subducting Cocos plate in central Mexico. J Geophys Res 115(B6):B06,310. https://doi.org/10.1029/2009JB006942

Kimura G, Hina S, Hamada Y, Kameda J, Tsuji T, Kinoshita M, Yamaguchi A (2012) Runaway slip to the trench due to rupture of highly pressurized megathrust beneath the middle trench slope: the tsunamigenesis of the 2011 Tohoku earthquake off the east coast of northern Japan. Earth Planet Sci Lett 339–340:32–45. https://doi.org/10.1016/j.epsl.2012.04.002

Kobayashi A (2014) A long-term slow slip event from 1996 to 1997 in the Kii Channel, Japan. Earth Planets Space 66(1):9. https://doi.org/10.1186/1880-5981-66-9

Kodaira S, Takahashi N, Nakanishi A, Miura S, Kaneda Y (2000) Subducted seamount imaged in the Rupture Zone of the 1946 Nankaido Earthquake. Science 289:104–106. https://doi.org/10.1126/science.289.5476.104

Kodaira S, Kurashimo E, Park JO, Takahashi N, Nakanishi A, Miura S, Iwasaki T, Hirata N, Ito K, Kaneda Y (2002) Structural factors controlling the rupture process of a megathrust earthquake at the Nankai trough seismogenic zone. Geophys J Int 149(3):815–835. https://doi.org/10.1046/j.1365-246X.2002.01691.x

Kodaira S, Iidaka T, Kato A, Park JO, Iwasaki T, Kaneda Y (2004) High pore fluid pressure may cause silent slip in the Nankai Trough. Science 304:1295–1298. https://doi.org/10.1126/science.1096535 (New York, NY)

Kodaira S, Hori T, Ito A, Miura S, Fujie G, Park JO, Baba T, Sakaguchi H, Kaneda Y (2006) A cause of rupture segmentation and synchronization in the Nankai trough revealed by seismic imaging and numerical simulation. J Geophys Res 111(B9):B09,301. https://doi.org/10.1029/2005JB004030

Kurashimo E, Iwasaki T, Iidaka T, Kato A, Yamazaki F, Miyashita K, Shibutani T, Ito K, Takeda T, Obara K, Hirata N (2013) Along-strike structural changes controlled by dehydration-related fluids within the Philippine Sea plate around the segment boundary of a megathrust earthquake beneath the Kii peninsula, southwest Japan. Geophys Res Lett 40(18):4839–4844. https://doi.org/10.1002/grl.50939

Li J, Shillington DJ, Bécel A, Nedimović MR, Webb SC, Saffer DM, Keranen KM, Kuehn H (2015) Downdip variations in seismic reflection character: implications for fault structure and seismogenic behavior in the Alaska subduction zone. J Geophys Res Solid Earth 120(11):7883–7904. https://doi.org/10.1002/2015JB012338

Miyazaki S, Heki K (2001) Crustal velocity field of southwest Japan: subduction and arcarc collision. J Geophys Res 106(B3):4305–4326. https://doi.org/10.1029/2000JB900312

Mochizuki K, Fujie G, Sato T, Kasahara J, Hino R, Shinohara M, Suyehiro K (1998) Heterogeneous crustal structure across a seismic block boundary along the nankai trough. Geophys Res Lett 25(13):2301–2304. https://doi.org/10.1029/98GL51867

Mochizuki K, Nakamura M, Kasahara J, Hino R, Nihisno M, Kuwano A, Nakamura Y, Yamada T, Shinohara M, Sato T, Moghaddam PP, Kanazawa T (2005) Intense PP reflection beneath the aseismic forearc slope of the Japan Trench subduction zone and its implication of aseismic slip subduction. J Geophys Res 110(B1):B01,302. https://doi.org/10.1029/2003JB002892

Mochizuki K, Nakahigashi K, Kuwano A, Yamada T, Shinohara M, Sakai S, Kanazawa T, Uehira K, Shimizu H (2010) Seismic characteristics around the fault segment boundary of historical great earthquakes along the Nankai Trough revealed by repeated long-term OBS observations. Geophys Res Lett 37(9):L09,304. https://doi.org/10.1029/2010GL042935

Nedimović MR, Hyndman RD, Ramachandran K, Spence GD (2003) Reflection signature of seismic and aseismic slip on the northern Cascadia subduction interface. Nature 424(6947):416–420. https://doi.org/10.1038/nature01840

Obara K, Hirose H, Yamamizu F, Kasahara K (2004) Episodic slow slip events accompanied by non-volcanic tremors in southwest Japan subduction zone. Geophys Res Lett 31(23):L23,602. https://doi.org/10.1029/2004GL020848

Okino K, Ohara Y, Kasuga S, Kato Y (1999) The Philippine Sea: new survey results reveal the structure and the history of the marginal basins. Geophys Res Lett 26(15):2287–2290. https://doi.org/10.1029/1999GL900537

Park JO, Naruse H, Bangs NL (2014) Along-strike variations in the Nankai shallow décollement properties and their implications for tsunami earthquake generation. Geophys Res Lett 41(20):7057–7064. https://doi.org/10.1002/2014GL061096

Peacock SM, Wang K (1999) Seismic consequences of warm versus cool subduction metamorphism: examples from southwest and northeast Japan. Science 286:937–939. https://doi.org/10.1126/science.286.5441.937

Reyners M, Eberhart-Phillips D (2009) Small earthquakes provide insight into plate coupling and fluid distribution in the Hikurangi subduction zone, New Zealand. Earth Planet Sci Lett 282(1–4):299–305. https://doi.org/10.1016/j.epsl.2009.03.034

Saffer DM, Tobin HJ (2011) Hydrogeology and mechanics of subduction zone forearcs: fluid flow and pore pressure. Annu Rev Earth Planet Sci 39(1):157–186. https://doi.org/10.1146/annurev-earth-040610-133408

Scholz C (1998) Earthquakes and friction laws. Nature 391:37–42. https://doi.org/10.1038/34097

Screaton E, Saffer D, Henry P, Hunze S (2002) Porosity loss within the underthrust sediments of the Nankai accretionary complex: implications for overpressures. Geology 30(1):19–22. https://doi.org/10.1130/0091-7613(2002)030<0019:PLWTUS>2.0.CO;2

Shelly DR, Beroza GC, Ide S, Nakamula S (2006) Low-frequency earthquakes in Shikoku, Japan, and their relationship to episodic tremor and slip. Nature 442:188–191. https://doi.org/10.1038/nature04931

Shelly DR, Beroza GC, Ide S (2007) Non-volcanic tremor and low-frequency earthquake swarms. Nature 446(7133):305–307. https://doi.org/10.1038/nature05666

Shiomi K, Park J (2008) Structural features of the subducting slab beneath the Kii Peninsula, central Japan: seismic evidence of slab segmentation, dehydration, and anisotropy. J Geophys Res 113(B10):B10,318. https://doi.org/10.1029/2007JB005535

Skarbek RM, Saffer DM (2009) Pore pressure development beneath the décollement at the Nankai subduction zone: implications for plate boundary fault strength and sediment dewatering. J Geophys Res Solid Earth 114(7):B07,401. https://doi.org/10.1029/2008JB006205

Song TRA, Helmberger DV (2007) Validating tomographic model with broad-band waveform modelling: an example from the LA RISTRA transect in the southwestern United States. Geophys J Int 171(1):244–258. https://doi.org/10.1111/j.1365-246X.2007.03508.x

Song TRA, Helmberger DV, Brudzinski MR, Clayton RW, Davis P, Perez-Campos X, Singh SK (2009) Subducting slab ultra-slow velocity layer coincident with silent earthquakes in southern Mexico. Science 330(5926):502–506. https://doi.org/10.1126/science.1167595

Tsuji Y, Nakajima J, Hasegawa A (2008) Tomographic evidence for hydrated oceanic crust of the Pacific slab beneath northeastern Japan: implications for water transportation in subduction zones. Geophys Res Lett 35(14):L14,308. https://doi.org/10.1029/2008GL034461

Zhao D, Huang Z, Umino N, Hasegawa A, Kanamori H (2011) Structural heterogeneity in the megathrust zone and mechanism of the 2011 Tohoku-oki earthquake (Mw 9.0). Geophys Res Lett 38(17):L17,308. https://doi.org/10.1029/2011GL048408

Chapter 2
Inverse Water-Layer Filter Method

Abstract Teleseismic waveforms recorded by ocean-bottom seismometers (OBSs) are affected by water reverberations. In this case, conventional methods, i.e., deconvolution of horizontal-component records by vertical-component records, fail to calculate receiver functions (RFs) correctly since the source wavelets cannot be approximated by the vertical-component records. We formulate a frequency filter termed the water-layer filter (WLF), which represents the impulse response of the water layer. With the inverse of the filter, we can remove the water reverberations from the vertical-component records. We demonstrate the effectiveness of the new method by conducting synthetic experiments.

Keywords Receiver function · Ocean-bottom seismometer
Water reverberations · Water-layer filter

2.1 Introduction

The seawater column traps seismic waves as they are reflected at both the sea surface and the seafloor; this is referred to as water reverberation. In the case of teleseismic and deep local events, water reverberations strongly affect the vertical-component records of the OBSs because these seismic waves propagate vertically to the seafloor as compressional waves in the water layer. Therefore, the waveforms of the vertical-component records are altered from their source wavelets. In such a case, deconvolution of the horizontal component by the vertical component fails to estimate the receiver functions (RFs) correctly.

In this chapter, we first propose a method for removing water reverberations from the vertical-component records. The key point of our strategy is to treat the water reverberation as a linear filter that acts on the incident wave to the water column from below the seafloor. We refer to this filter as a water-layer filter (WLF). The same concept can be found in marine reflection seismology; to the best of our knowledge, it first appeared in the 1950s (Backus 1959). In the latter half of this chapter, we demonstrate how our method can improve RF estimation by conducting synthetic tests.

© Springer Nature Singapore Pte Ltd. 2018

17

T. Akuhara, *Fluid Distribution Along the Nankai-Trough Megathrust Fault off the Kii Peninsula*, Springer Theses, https://doi.org/10.1007/978-981-10-8174-3_2

2.2 Definition of Water-Layer Filter

Let us consider an infinite half-space overridden by a water layer with an OBS on the seafloor (Fig. 2.1a). If an impulsive plane P-wave with unit amplitude enters the water layer from below, the wave repeatedly reflects on the liquid–solid boundary and at the top of the water layer. In such a situation, we may express the OBS vertical-component record, $w(t)$, as follows:

$$w(t) = \{(1 + R)\cos\theta_P - R_{P-SV}\sin\theta_S\}\delta(t)$$
$$- \sum_{n=1}^{\infty}(-1)^n T(T'\cos\theta_P - T'_{P-SV}\sin\theta_S)R^{n-1}\delta(t - n\tau) \qquad (2.1)$$

and

$$\tau = \frac{2h}{c}\sqrt{1 - c^2 p^2}, \qquad (2.2)$$

where R and R_{P-SV} represent P-to-P and P-to-SV reflection coefficients, respectively, at the liquid–solid boundary of the seafloor. The upward P-to-P, downward P-to-P, and downward P-to-SV transmission coefficients at the boundary are denoted by T, T', and T'_{P-SV}, respectively. The incident angle of the P-wave and the reflection angle of the S-wave below the seafloor are represented by θ_P and θ_S, respectively. Here, we formulated Eq. 2.1 such that the signs of all reflection and transmission coefficients are positive. The first term on the right-hand side of Eq. 2.1 corresponds to the direct arrival, including the downward reflection on the seafloor, and the second term denotes the successive nth reverberations. In Eq. 2.2, τ represents the two-way travel time within the water layer, which is expressed by the wave speed in water, c, water-layer thickness, h, and ray parameter, p.

If we assume the ray parameter to be zero, equivalent to the vertical incidence, we can obtain approximate forms of Eqs. 2.1 and 2.2:

$$w(t; \tau, R) = (1 + R)\delta(t) - \sum_{n=1}^{\infty}(-1)^n(1 - R^2)R^{n-1}\delta(t - n\tau) \qquad (2.3)$$

and

$$\tau = \frac{2h}{c}. \qquad (2.4)$$

Here, we employed the following five approximations related to the assumption of the vertical incidence: $c^2 p^2 \sim 0$, $R_{P-SV}\sin\theta_S \sim 0$, $T'_{P-SV}\sin\theta_S \sim 0$, $\cos\theta_P \sim 1$, and $TT' \sim 1 - R^2$. Beneath the sea floor, an unconsolidated sediment layer with low seismic velocity allows the incident angle to be almost zero, so these approximations are reasonable. We demonstrated this by theoretical calculations based on the Zoeppritz equations, assuming typical physical properties for a sediment (Vp = 2.0 km/s, Vs = 0.5 km/s, and a density of 1.8 g/cm³ (Hamilton 1979, 1978) and a seawater

(a)

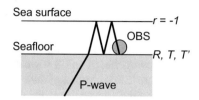

(b)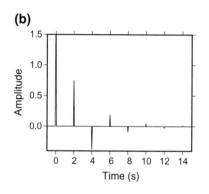

(c)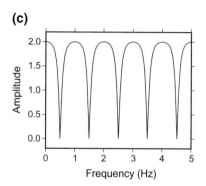

Fig. 2.1 **a** Schematic illustration of water reverberations, **b** time domain representation and **c** amplitude spectrum of a water-layer filter ($\tau = 1.0$ s and $R = 0.5$). Modified from Akuhara and Mochizuki (2015) by permission of John Wiley & Sons Ltd

Table 2.1 Quantitative verification of approximations used in Eqs. 2.3 and 2.4

	$c^2 p^2$	$R_{P-SV} \sin \theta_S$	$T'_{P-SV} \sin \theta_S$	$1 - R^2 - TT'$	$\cos \theta_P$
Approximated value	0	0	0	0	1
Theoretical value	8×10^{-3}	2×10^{-3}	1×10^{-3}	2×10^{-2}	0.993

layer ($c = 1.5$ km/s and a density of 1.5 g/cm^3). We employed a ray parameter of 0.06 s·km^{-1} for this calculation. The results show that these five approximations hold well (Table 2.1).

Next, we recast Eq. 2.3 in the frequency domain:

$$W(\omega; \tau, R) = \left(R - \frac{1}{R} \right) \frac{1}{1 + R \exp(-i\omega\tau)} + \frac{1+R}{R}. \tag{2.5}$$

We define the WLF using Eq. 2.3 or 2.5, which corresponds to a transfer function producing water reverberations. The application of the inverse WLF (IWLF), or deconvolution by the WLF, eliminates water reverberations. The amplitude spectra of the WLF fall strictly to zeros, irrespective of R values, at each constant frequency interval of $1/\tau$ Hz (Fig. 2.1c). In practice, these zeros make the deconvolution operation unstable, requiring damping factors such as water-level damping (Clayton and Wiggins 1976). Hereafter, the application of the IWLF indicates a spectral division by Eq. 2.5, with a water-level damping of 0.05. We arrived at this damping value by trial and error.

Alternative methods to remove water reverberations have been developed, which decompose observed wave fields into upgoing and downgoing p- and S-wave components (Bostock and Trehu 2012; Thorwart and Dahm 2005). Although their early methods require hydro-pressure records in addition to displacement data, the recent method proposed by Bostock and Trehu (2012) requires only displacement data. Unlike our method, their method enables the treatment of nonzero ray parameters, although it also requires a number of tuning parameters (e.g., density and p- and S-wave velocities of the seafloor and the water depth). Our method corresponds to the special case of their method with zero ray parameters. The marked advantages of our method are its easy implementation and the reduced number of unknown parameters: Our method demands only a single operation by a filter expressed with two unknown parameters, τ and R.

2.3 Synthetic Test

2.3.1 Methodology

To confirm the efficiency of the IWLF, we conducted synthetic tests with two different 1D layered models (Table 2.2). The first model, Model A, is composed of water, crust, and mantle layers. The second model, Model B, possesses an additional thin sediment layer beneath the water layer, so is more realistic than Model A. In our synthetic test both vertical and radial-component records were first computed by a propagator matrix method (Haskell 1953; Kumar et al. 2011). We synthesized these waveforms with a ray parameter of 0.06 s·km^{-1}, corresponding to oblique incidence. Then, we applied IWLF to the vertical-component record to remove water reverberations. Finally, we estimated radial RF by deconvolving the radial-component record with the filtered vertical-component record. We performed this deconvolution using the extended-time multitaper correlation (ETMTC) method (Shibutani et al. 2008). Resultant RFs were low-pass filtered to 4.0 Hz.

For the application of IWLF, we used τ and R values calculated as follows:

$$\tau = \frac{2h_1}{\alpha_1} \tag{2.6}$$

and

$$R = \frac{\rho_1\alpha_1 - \rho_2\alpha_2}{\rho_1\alpha_1 + \rho_2\alpha_2}. \tag{2.7}$$

In these equations, α_1, ρ_1, and h_1 represent P-wave velocity, density, and thickness of the water layer, respectively. Similarly, α_2 and ρ_2 represent P-wave velocity and density of the topmost layer beneath the seafloor, respectively. In addition to these theoretical values, we also tested the slightly shifted values to understand how poorly constrained parameters influence RF estimation in practical applications. The shift values we employed were 0.2 for R and 0.05 s for τ.

2.3.2 *Results and Discussion*

Figure 2.2 shows the effect of IWLF on the synthetic vertical component records. In both Model A and B, the original waveforms show the first water reverberations 2.1 s after the direct P arrivals. The amplitude of this first water reverberation is much larger for Model B than for Model A. In Model B, a small impedance contrast across the seafloor allows more energy to enter the seawater, making the first water reverberation stronger. The second water reverberation of Model B, which arrives 4.2 s after the direct P phase with negative polarity, is much weaker than the first reverberation. This is because the most of the downgoing wave energy in the water layer passes through the seafloor due to the small impedance contrast.

The waveform of Model B shows a dominant negative arrival around 3 s after the direct P arrival. This arrival corresponds to the phase that is first reflected at the sea surface and then at the bottom of the sediment layer. Although this phase may have the potential to distort RFs as well as water reverberations, such a phase seems less dominant in actual observation data than expected here (see Fig. 1.2, where only the first water reverberation seems prominent after the direct P arrival). The actual subsurface structure might possess weaker impedance at the bottom of the sediment layer than that in Model B.

Red traces in Fig. 2.2 represent the vertical-component records processed with the IWLF; note that they are normalized so that the amplitudes of the direct P arrivals are

Table 2.2 Two types of layered model used in synthetic tests

	Layer type	Thickness (km)	Vp (km/s)	Vs (km/s)	Density (g/cm^3)	Ray parameter (s·km^{-1})
Model A	Water	1.6	1.5	–	1.0	0.06
	Crust	20.0	6.0	3.5	2.7	
	Mantle	–	8.1	4.7	3.4	
Model B	Water	1.6	1.5	–	1.0	0.06
	Sediment	0.7	1.0	0.4	1.8	
	Crust	20.0	6.0	3.5	2.7	
	Mantle	–	8.1	4.7	3.4	

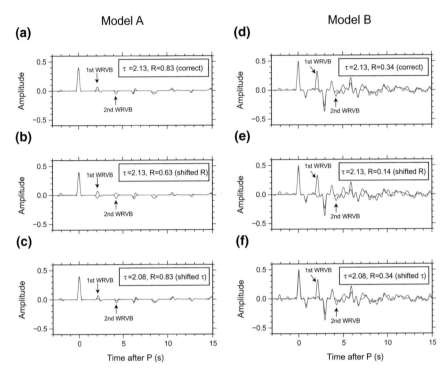

Fig. 2.2 Synthetic vertical-component records calculated with (**a**–**c**) Model A and (**d**–**f**) Model B. Gray waveforms represent vertical-component records before the application of the inverse water-layer filter. Red waveforms are the filtered waveforms with (**a, d**) the correct parameters, and slightly shifted (**b, e**) R and (**c, f**) τ values. All waveforms are low-pass filtered below 4.0 Hz. Note that the amplitudes of the inverse water-layer filtered waveforms are multiplied by $1 + R$ to compensate for the amplitude reduction caused by the removal of downward reflections at the seafloor. WRVB = water reverberation

equal to those of the original waveforms for comparison. We find that water reverberations are successfully removed, or at least effectively suppressed, by the IWLF, even if we use τ and R values that are shifted from their true values (Figs. 2.2b, c, e, and f). We also notice that the application of the IWLF produces acausal signals before the P-wave arrival; however, these signals are trivial compared with the dominant water reverberations on the original waveforms.

The synthetic RFs from the original and the inverse water-layer filtered vertical-component records are presented in Fig. 2.3. Note that the RFs processed with the IWLF are normalized so that the amplitudes of the direct P arrivals are equal to those of the original RFs. In the case of Model A, the original RF exhibits (i) a positive P-to-S conversion phase from the Moho (Ps_M in Fig. 2.4) at 2.5 s after the direct P arrival, and (ii) artificial peaks due to water reverberations at 2.1 s and 4.3 s. These artificial peaks are successfully removed, or at least reduced after the application of IWLF, indicating the efficiency of our method.

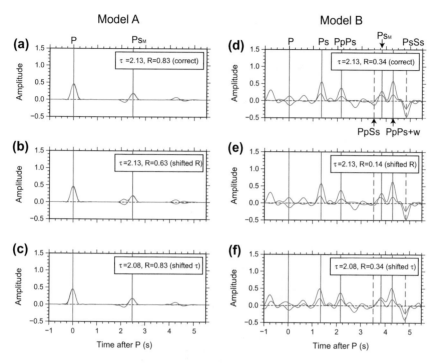

Fig. 2.3 Synthetic receiver functions (RFs) calculated with (**a–c**) Model A and (**d–f**) Model B. In each panel, red and gray traces are synthetic RFs calculated with and without removal of water reverberations, respectively. Note that the red traces are normalized so that the amplitudes of the direct P arrivals are equal to those of gray traces. All RFs are low-pass filtered to 4.0 Hz. Solid and dashed blue lines show expected timings of positive and negative phases, respectively. The definition of the phases appearing in this figure are illustrated in Fig. 2.4

More drastic improvement can be seen in the case of Model B. The original RF shows many meaningless peaks throughout the records (meaningless in this sense that the peak cannot be explained from by the model). For example, there is no reasonable explanation for the positive phase at ~0.6 s and negative phase at ~2.7 s. These artificial peaks are considered to be generated by strong water reverberations on the vertical component. On the other hand, the RF processed with IWLF shows six isolated phases after the direct P arrival. With this improved RF, we can interpret all the phases as (i) P-to-S conversion phase at the Moho (Ps_M) and the bottom of the sediment layer (Ps), and (ii) sediment-related reverberations (PpPs, PpSs, PsSs, and PpPs+w, see Fig. 2.4b).

The RF calculated with the IWLF reproduces the amplitudes (not only the timings) of these phases, to some degree. We confirmed this by comparing the RFs with the synthesized radial-component waveforms (Fig. 2.5).

For both models, the artificial peaks on the RFs were suppressed even if we employed shifted values (Figs. 2.3b, c, e, and f). Determining the thresholds of an

(a) Common in Model A and B

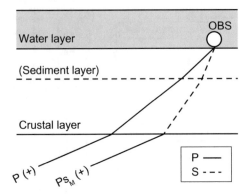

(b) Model B only

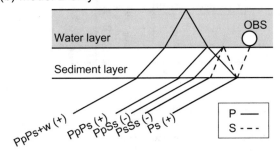

Fig. 2.4 Schematic illustration of major phases appearing in synthetic receiver functions (Fig. 2.3). The sign following the phase names represents their polarity in the radial receiver function

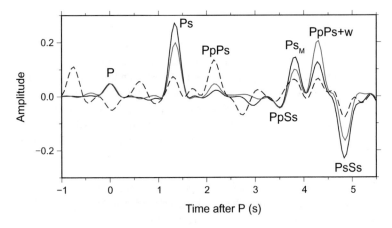

Fig. 2.5 Synthesized radial-component record (solid black line), radial RF estimated with the removal of water reverberations (red line), and radial RF estimated by an ordinary method (dashed black line). All traces were calculated with Model B and are low-pass filtered to 4.0 Hz

acceptable shift amount is difficult because it depends on model structure and the frequency range of low- or band-pass filters. From our experience the acceptable shift amounts are typically about 0.1 s and 0.3 for τ and R, respectively. In Chap. 3, it is shown that based on observed data, we can constrain the tau and R values within these ranges.

References

Akuhara T, Mochizuki K (2015) Hydrous state of the subducting philippine sea plate inferred from receiver function image using onshore and offshore data. J Geophys Res Solid Earth 120(12):8461–8477. https://doi.org/10.1002/2015JB012336

Backus MM (1959) Water reverberations—Their nautre and elimination. Geophysics 24(2):233–261. https://doi.org/10.1190/1.1438579

Bostock MG, Trehu AM (2012) Wave-field decomposition of ocean bottom seismograms. Bull Seismol Soc Am 102(4):1681–1692. https://doi.org/10.1785/0120110162

Clayton RW, Wiggins RA (1976) Source shape estimation and deconvolution of teleseismic body-waves. Geophys J Int 47(1):151–177. https://doi.org/10.1111/j.1365-246X.1976.tb01267.x

Hamilton EL (1978) Sound velocitydensity relations in sea-floor sediments and rocks. J Acoust Soc Am 63(2):366–377. https://doi.org/10.1121/1.381747

Hamilton EL (1979) Vp/Vs and Poisson's ratios in marine sediments and rocks. J Acoust Soc Am 66(4):1093–1101. https://doi.org/10.1121/1.383344

Haskell NA (1953) The dispersion of surface waves on multilayered media. Bull Seismol Soc Am 43(1):17–34

Kumar P, Kawakatsu H, Shinohara M, Kanazawa T, Araki E, Suyehiro K (2011) P and S receiver function analysis of seafloor borehole broadband seismic data. J Geophys Res Solid Earth 116(12):1–17. https://doi.org/10.1029/2011JB008506

Shibutani T, Ueno T, Hirahara K (2008) Improvement in the extended-time multitaper receiver function estimation technique. Bull Seismol Soc Am 98(2):812–816. https://doi.org/10.1785/0120070226

Thorwart M, Dahm T (2005) Wavefield decomposition for passive ocean bottom seismological data. Geophys J Int 163(2):611–621. https://doi.org/10.1111/j.1365-246X.2005.02761.x

Chapter 3
Application of Inverse Water-Layer Filter Method

Abstract Here, we address several issues regarding the application of an inverse water-layer filter (IWLF) method to observed data. Initially, we present a nonlinear inversion method to determine IWLF parameters. We use vertical components of tele-seismic P waveforms as inputs for the inversion and simultaneously solve for a source wavelet, two-way travel times within the water layer, and reflection coefficients on the seafloor. A simulated annealing algorithm is employed for this optimization. We then investigate the validity of the IWLF method using autocorrelation functions. Finally, we observe that radial receiver functions calculated using the IWLF method show clearer phases than those calculated by a typical method.

Keywords Receiver function · Ocean-bottom seismometer
Water reverberations · Nonlinear inversion

3.1 Introduction

In Chap. 2, we introduced the inverse water-layer filter (IWLF) method to remove water reverberations from vertical-component records of teleseismic P-waves and demonstrated its effectiveness in receiver function (RF) estimation with synthetic experiments. In this chapter, we address several issues relating to its application to real observation data. The first compulsory task is the determination of the IWLF parameters, τ and R, from observed data. Although there should be various ways to achieve this goal, we introduce a method involving the nonlinear waveform inversion in Sect. 3.2. A major advantage of this inversion method is that no additional data is required other than ocean-bottom seismometer (OBS) vertical-component records.

The second task we should address is evaluation of the effectiveness of the IWLF method with real data (Sect. 3.3). Our formulation of the IWLF depends on two assumptions, the vertical incidence of seismic waves and the flatness of the seafloor. Therefore, we must pay careful attention to whether the IWLF works well with real data.

© Springer Nature Singapore Pte Ltd. 2018
T. Akuhara, *Fluid Distribution Along the Nankai-Trough Megathrust Fault off the Kii Peninsula*, Springer Theses, https://doi.org/10.1007/978-981-10-8174-3_3

At the end of this chapter, we show a brief example of radial RFs estimated with the removal of water reverberations (Sect. 3.4). The resultant RFs exhibit more distinct phases than those calculated by a typical method.

3.2 Inversion Analysis for Water-Layer Filter Parameters

3.2.1 Overview of Inversion Analysis

Let us consider a teleseismic P-wave recorded at a densely deployed array of OBS stations. We may express the vertical-component records of the ith OBS, $u^i(t)$, as the convolution of a source wavelet including effects from a near-source structure, $s(t)$, and a vertical-component impulse response of a near-receiver structure, $G_V^i(t)$, after correcting the instrumental response:

$$u^i(t) = s(t) * G_V^i(t). \tag{3.1}$$

The asterisk in Eq. 3.1 denotes the convolution operation. If we further assume that the vertical impulse response is equal to the response of the water layer, namely $w^i(t; \tau^i, R^i)$, we can rewrite Eq. 3.1 as follows:

$$u^i(t) = s(t) * w^i(t; \tau^i, R^i). \tag{3.2}$$

In general, separating the two terms in the right-hand side of Eq. 3.2 from a single event record is difficult; there is a trade-off between estimation of a source wavelet and deciding on the water-layer filter (WLF) parameters. Fortunately, we can resolve this trade-off by using multiple OBS records because the source wavelet is common among all OBS stations whereas the impulse responses are not. To achieve this separation, we perform a nonlinear waveform inversion using a simulated annealing (SA) algorithm (Kirkpatrick et al. 1983).

Such inversion analyses using the SA algorithm have been previously conducted to extract source wavelets and/or relative arrival times (Chevrot 2002; Iritani et al. 2010, 2014; Tonegawa et al. 2013). The application of the SA method to our case is highly motivated by the work of Iritani et al. (2010, 2014), who successfully identified the arrivals of core phases even if they were overlapped by the long tails of earlier phases. Tonegawa et al. (2013) estimated source wavelets using the SA algorithm and computed RFs from the deconvolution using the estimated source wavelets. The method of Tonegawa et al. (2013) is an alternative to our IWLF method; however, we prefer our IWLF method because there is no necessity to estimate the source wavelets of all events. This is a distinct advantage in the case of OBS, because their relatively noisy data does not allow us to successfully estimate the source wavelets of all events.

We use vertical-component records of a single teleseismic or deep event recorded by multiple OBS stations as inputs to the inversion problem. Parameters to be estimated are the source wavelet, $s(t)$, the arrival times of the direct P-wave, t_p^i, the normalization factor,[1] a^i, and the two WLF parameters, τ^i and R^i. The superscripts of these parameters, i, represent indices discriminating OBSs. The synthetic waveforms of the ith station, $u_{syn}^i(t)$, can then be represented as follows:

$$u_{syn}^i(t) = a^i \cdot s(t) * \delta(t - t_p^i) * w(t; \tau^i, R^i) \tag{3.3}$$

If we have N_{sta} stations and N_{smp} samples for a source wavelet, then total number of model parameters amounts to $N_{smp} + N_{sta} \times 4$.

3.2.2 Prior Estimation of Phase Arrivals

Cycle skipping, due to false determination of t_p by a few multiples of a dominant wave period, is one of the major factors making the inversion unsuccessful. A good strategy to avoid this problem is to make the best prior estimation for t_p, so that we limit its search range during inversion. Thus, we performed automatic phase picking using short-term average to long-term average ratio algorithms (e.g., Abt et al. 2010) for the initial estimation of t_p. We briefly introduce the method below.

We first applied a band-pass filter of 0.1–2.0 Hz to the vertical component of teleseismic or deep event records satisfying the prerequisite condition described in Sect. 1.4.2 (Fig. 3.1a). Then, we calculated their envelope functions using the Hilbert transform (Fig. 3.1b). Next, we shifted 1-s and 4-s time windows over the envelope functions to evaluate the signal-to-noise ratio (SNR) function (Fig. 3.1c). The SNR function was calculated as the root-mean-square amplitude ratio of the two time windows. The moving range was ± 10 s around the theoretical P-wave arrival times based on the IASP91 model (Kennett and Engdahl 1991). Finally, we selected the time point with the maximum SNR value as the initial estimate of t_p.

In the following inversion analysis, we use the band-pass filtered waveforms with maximum SNR values of > 3.0 as the input data. The length of the input waveform is 13 s, starting from 3 s before the prior estimation of t_p.

---

[1]This normalization factor is introduced to compensate for variation in observed amplitudes of the direct arrivals among stations. This factor may not be necessary if we attribute the amplitude variation to the difference in reflection coefficients at the seafloor. Indeed, similar results are obtained without the normalization factor (Akuhara et al. 2016).

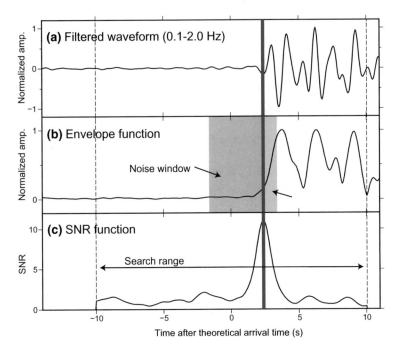

Fig. 3.1 Process of automatic arrival time selection. We first (**a**) applied a band-pass filter (0.1–2.0 Hz) to the vertical-component event record, then (**b**) calculated the envelope function, and finally (**c**) searched for the maximum signal-to-noise ratio (SNR) function by moving the time window. The SNR functions were computed using the envelope function. The blue vertical bar represents the resultant arrival time selection. The pink and light green areas denote the time windows for noise and signal, respectively, to compute the SNR function

3.2.3   Inversion Scheme

We performed waveform inversion for each event whose waveform records show a SNR > 3.0 at eight or more OBSs. We normalized all input data by the maximum amplitude among all OBS records. Our inversion involved 2000 iterations to determine an optimum solution. The single iteration consists of three processes: random sample generation, misfit measurement, and evaluation processes.

Random Sample Generation

At each iteration, we randomly generate a set of model parameters, a^i, R^i, t_p^i, τ^i, and s_j, under the following conditions:

$$0.3 \leq a^i \leq 1.0, \tag{3.4}$$

$$0.1 \leq R^i \leq 1.0, \tag{3.5}$$

$$t_p^{'i} - 0.5(s) \leq t_p^i \leq t_p^{'i} + 0.5(s), \tag{3.6}$$

$$\tau^{'i} - 0.25(s) \leq \tau^i \leq \tau^{'i} + 0.25(s), \tag{3.7}$$

and

$$s_j = s_j' \pm 0.01, \tag{3.8}$$

where s_j is the jth discrete time sample of source wavelets and the prime notations (such as $t_p^{'i}$, $\tau^{'i}$, and s_j') represent the corresponding parameters at the previous iteration step. At the initial iteration, t_p^i, τ^i, and s_j were fixed to the prior estimation (Sect. 3.2.2), the expected value from OBS depth, and zero, respectively. The OBS depths were measured by acoustic ranging on installed or recovering OBSs. We set the source duration to 10 s for all events, which corresponds to N_{smp}= 200 in our case with a 20 Hz sampling rate.

Misfit measurement

Each time a trial parameter was generated, we calculated the misfit between the observed and synthetic waveforms, m, as the summation of the L1-norm difference over all records:

$$m = \sum_{i=1}^{N_{sta}} \sum_j |u_{syn}^i(j \Delta t) - u_{obs}^i(j \Delta t)|, \tag{3.9}$$

where u_{obs}^i represents the observed waveform at the ith station and Δt represents the sampling interval.

Evaluation

The newly generated parameter is accepted as the optimum solution if the misfit value decreases appropriately. For t_p and τ, we accepted the trial parameter with a certain probability even if the misfit became larger. This probability was expected to become smaller as the iteration number increased as follows

$$probability = \exp\left(-\frac{m - m'}{T_0 \cdot \gamma^k}\right), \tag{3.10}$$

where k is the iteration number and m' is a misfit value measured by model parameters from the previous step. In addition, T_0 and γ are tuning parameters of the

simulated annealing algorithm referred to as initial temperature and cooling sched-
ule, respectively. We fixed T_0 at three times the initial misfit value and γ at 0.99 in
accordance with Iritani et al. (2010).

3.2.4 Results and Discussion

The inversion results show good reproduction of the observed waveforms even when
the onset of water reverberation is contaminated by the coda of the direct arrivals
(Fig. 3.2). For the following statistical analysis, we evaluated the quality of the final
solutions by calculating cross-correlation coefficients (CCs) between the synthetic
and observed waveforms at each OBS, and discarded ill-fit seismograms with CC
< 0.8. In addition, if the inversion failed to produce eight or more seismograms with
CC ≥ 0.8, we discarded all results from the inversion (Fig. 3.3).

To confirm the robustness, we repeated this inversion eight times for each event
using different random number sequences. We then calculated the averages and
standard deviations of τ and R across these eight inversions. As a result, we obtained
standard deviations of <0.05 s for τ and <0.1 for R for most events (Fig. 3.4). These
values are considerably smaller than the search ranges, suggesting stability in the
analysis. In the following section, we show the averaged τ^i and R^i values of the eight
estimations, rather than individual results.

Figure 3.5 shows the resultant τ and R values estimated from different events at
three OBSs: LS11, LS12, and LS20. In all cases, individual τ and R values were
within ±0.05 s and ±0.2, respectively, around their averaged values. If this scatter-
ing originated from the dipping seafloor, there should be a correlation between the
estimated τ (or R) values and the locations of the P-wave incidence into the water
column. However, we could not find such a correlation (Figs. 3.5b and c). We therefore postulate that the scattered τ and R distributions mainly reflected random errors;
therefore, we decided to use the averaged τ and R values of all events as parameters
for the IWLFs (i.e., stars in Fig. 3.5a).

As expected, the resulting station-averaged τ values are consistent with water
depth (Fig. 3.6a). Figure 3.6b shows a histogram of the deviations between the averaged τ and expected values (i.e., double the OBS depth divided by a typical wave
speed of 1.5 km/s). Although the peak is located at the center, we observed a relatively
large difference (>0.15 s) at several OBSs. Our previous synthetic tests (Sect. 2.3)
demonstrated such a large difference in τ values could disable the IWLF method.
Therefore, we conclude that estimating τ values based only on OBS depths would
be problematic in some cases.

Averaged R values for each OBS vary between 0.1 and 0.5, where smaller values
represent softer material with small P-wave velocity beneath the seafloor (Fig. 3.7).
The averaged value of all OBSs is 0.30, which is reasonable for typical sediments.
A weak positive correlation can be seen between the reflection coefficients and the

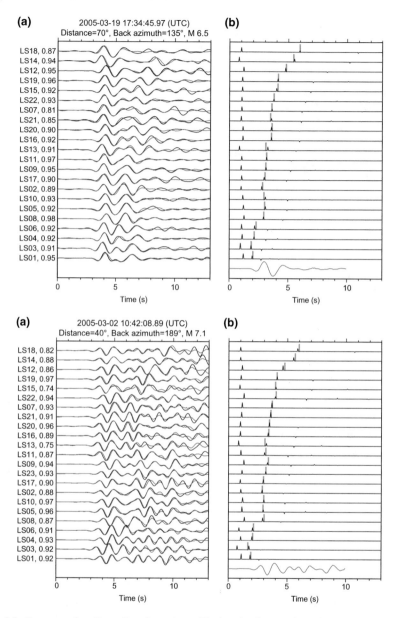

Fig. 3.2 Two examples of waveform inversion. **a** Black and red traces show observed and synthetic waveforms, respectively. OBS names and cross-correlation coefficients between the observed and synthetic waveforms are listed on the vertical axis. **b** Red waveform at the bottom shows a source wavelet, $s(t)$, determined by the inversion. Black pulses show WLFs convolved with $a^i \delta(t - t_p^i)$ in Eq. 3.3. Green bars represent the expected timing of the first reverberation inferred from the OBS depths

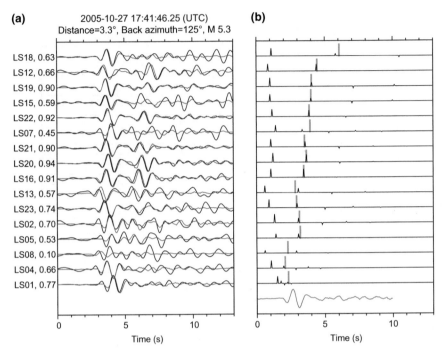

Fig. 3.3 An example of waveform inversion showing erroneous results. Note that all results from this inversion were discarded. All notations are the same as Fig. 3.2

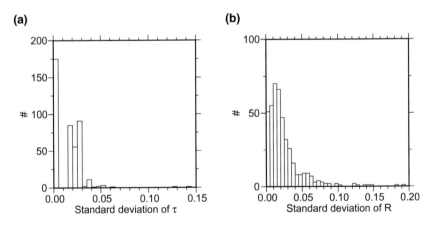

Fig. 3.4 Histograms of the standard deviations of **a** τ and **b** R over eight inversions using different random number sequences

OBS depths: smaller R values at shallower depths and larger R values at deeper depths. This tendency may represent some regionality of the sediment properties, but we avoid further interpretation of this phenomenon in this thesis.

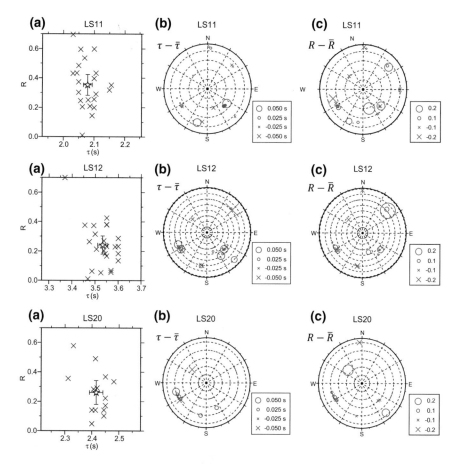

Fig. 3.5 Results of waveform inversion for three OBSs: LS11, LS12, and LS20. **a** Crosses represent τ and R values estimated from different events. Stars are their averaged values with 2σ standard errors. **b, c** Deviations of **b** τ and **c** R values from their average values against their back azimuths and horizontal propagation distances of the first water reverberation in the water column. The sizes of the crosses and circles indicate the extent of the deviations, and their locations represent the incident points of the first water reverberation into the seawater. The center of each polar coordinate is identical to the OBS location, and dashed circles denote intervals of 100 m

3.3 Quantitative Assessment of Inverse Water-Layer Filter

Seismograms with a short source duration help us confirm the efficiency of our IWLF method, because their water reverberations are clearly separated from the direct P-wave arrivals. Figure 3.8 shows an example of such seismograms for an OBS vertical component (black curves) compared with those after application of the IWLFs (red curves). In these examples, it can be seen that water reverberations are effectively

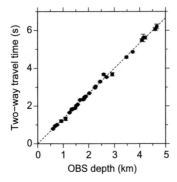 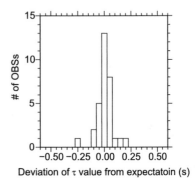

Fig. 3.6 **a** Correlation diagram between OBS depth and averaged τ value for each OBS. Error bars denote the 2σ confidence region of standard errors. Dashed line represents the expected two-way travel time from OBS depth, where we assumed a constant P-wave speed of 1.5 km/s within the water-layer and vertical incidence of seismic waves. **b** Histogram of differences between expected τ values from OBS depth and the estimated τ values from the inversion analysis

Fig. 3.7 Correlation diagram between OBS depth and averaged R value for each OBS. Error bars denote the 2σ confidence region of standard errors

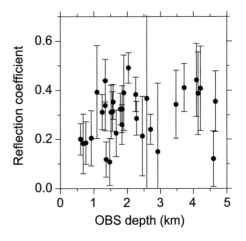

reduced or completely removed. The amplitudes of the direct arrivals are also slightly reduced due to the removal of downward reflection at the seafloor.

In more general cases, water reverberations are contaminated by the coda of direct P-wave arrivals; therefore, we cannot visually assess the efficiency of the IWLF. We further investigate the effectiveness by inspecting autocorrelation functions (ACFs) of the vertical-component records. Figure 3.9a shows the ACFs of the vertical-component records at LS11, LS12, and LS20. All event records processed here satisfy the prerequisite criteria described in Sect. 1.4. Clear and consistent positive peaks appear at around 2–4 s for all OBS stations, indicating water reverberations (Fig. 3.9a). It can be seen that these positive peaks are suppressed by applying the IWLF (Fig. 3.9b).

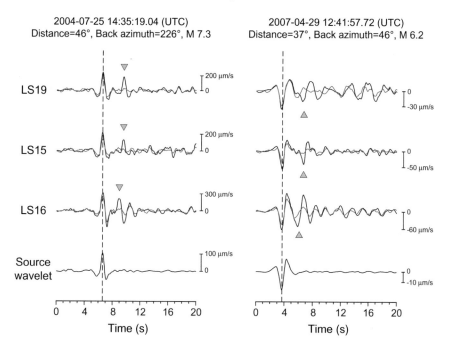

Fig. 3.8 Comparisons between original and inverse water-layer filtered seismograms. In the first to third rows, black and red waveforms show original and inverse water-layer filtered ocean-bottom seismometer vertical-component records, respectively. Waveforms in the fourth row indicate stacked traces of on-land records, which we regard as source wavelets. Dashed vertical lines denote the arrival times of large amplitudes, which were aligned by the maximum amplitude of the cross-correlation function. Green triangles represent the expected arrival times of the first water reverberations corresponding to the large amplitudes. Modified from Akuhara and Mochizuki (2015) by permission of John Wiley & Sons Ltd.

To quantify the effectiveness of the IWLF for each event record, we compared the difference in the root-mean-square amplitude of the ACFs before and after processing with the IWLFs (hereafter, we refer to this measurements as $\Delta_{RMS-ACF}$). Roughly speaking, $\Delta_{RMS-ACF}$ tends to be negative for most traces (Fig. 3.9c), which we believe stems from the reduction of water reverberations. If we use an incorrect τ value that is shifted by 0.15 s from the estimated value, IWLF produces artificial peaks and positive $\Delta_{RMS-ACF}$ values in a larger number traces (Fig. 3.10). Therefore, in the following analysis, we decided to use a negative $\Delta_{RMS-ACF}$ as a criterion against which to judge whether or not water reverberations are successfully removed. We refer to this as the "$\Delta_{RMS-ACF}$ criterion," hereafter.

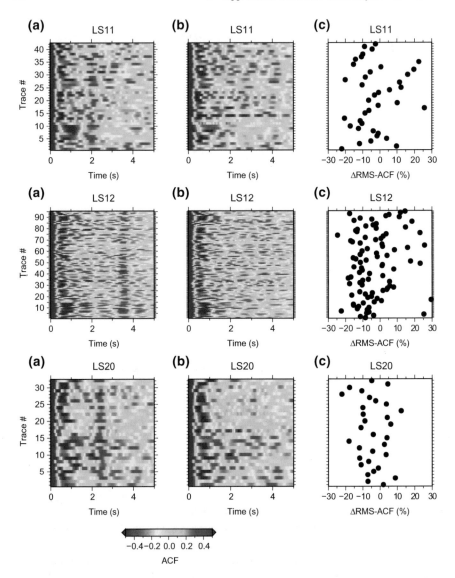

Fig. 3.9 Autocorrelation functions of vertical-component records **a** before and **b** after the application of inverse water-layer filters. **c** Differences in RMS amplitudes of the autocorrelation functions between before and after application of the inverse water-layer filter ($\Delta_{RMS-ACF}$)

3.4 Application to Receiver Function Estimation

In Fig. 3.11, we show radial RFs calculated at a single OBS station (LS11) by two different methods: the deconvolution of radial-component records with (1) original vertical-component records (Fig. 3.11a) and (2) vertical-component records pro-

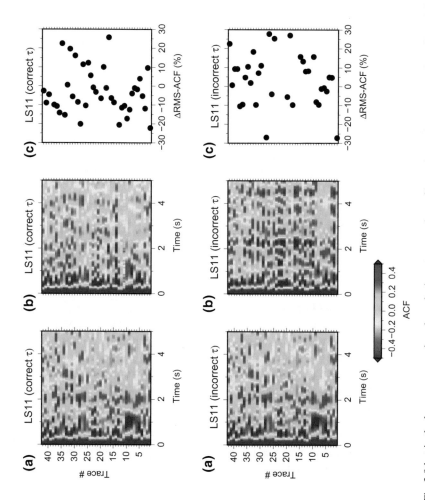

Fig. 3.10 The same as Fig. 3.9 but in the lower row, we show how the inverse water-layer filter using an incorrect τ value affects ACF and $\Delta_{RMS-ACF}$. The upper row indicates the results using the correct τ value

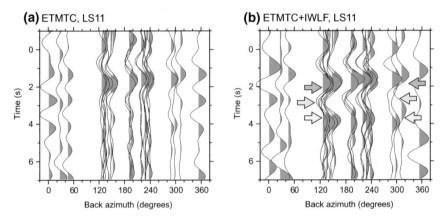

Fig. 3.11 Radial receiver functions (RFs) calculated at a single ocean-bottom seismometer (LS11): **a** without removal of water reverberations and **b** after removal of water reverberations. Positive amplitudes of RFs are shaded dark gray. Gray, pink, and pale blue arrows indicate major RF peaks, which we interpret as P-to-S conversion phases from the bottom of the sediment layer, plate interface, and oceanic Moho, respectively. All RFs are band-pass filtered between 0.1 and 1.0 Hz. Modified from Akuhara and Mochizuki (2015) by permission of John Wiley & Sons Ltd.

cessed with the IWLF (Fig. 3.11b). In both cases, the ETMTC method was employed to compute the RFs, and the resultant RFs were band-pass filtered between 0.1 and 1.0 Hz. The latter method employs two deconvolutions: the first to remove water reverberations and the second to calculate RFs, which seems to make the estimation unstable. Alternative method to avoid the second deconvolution is to apply the WLF to the radial component instead of the IWLF to the vertical component; however, we found that the results did not change significantly when using this modified method. Note that in Fig. 3.11, we only show RFs with vertical-component records satisfying the $\Delta_{RMS-ACF}$ criterion (Sect. 3.3).

Compared with the conventional method, our method produced clearer positive phases at around 4 s on RFs with back azimuths between 120° and 300° (pale blue arrows in Fig. 3.11). Based on a comparison with a tomography model (Akuhara et al. 2013), we interpreted these enhanced phases as the P-to-S conversion phases from the oceanic Moho of the subducting PHS plate. In addition, we can identify the P-to-S conversion phases from the top of the PHS plate and the bottom of the sediment layer (pink and gray arrows in Fig. 3.11, respectively). The arrival of the P-to-S conversion phases from the oceanic Moho and the plate interface seems delayed for other back azimuth ranges due to the dipping subducting slab, although the small number of events to the north makes it difficult to identify such moveout with confidence. Unlike usual RFs from on-land station data, RFs from OBS data have a less dominant signal at zero delay time corresponding to direct P-wave arrivals. This is because significantly reduced velocity in the unconsolidated sediment beneath the seafloor leads to a nearly vertical incidence angle of the incoming P-waves.

References

Abt DL, Fischer KM, French SW, Ford HA, Yuan H, Romanowicz B (2010) North American lithospheric discontinuity structure imaged by Ps and Sp receiver functions. J Geophys Res 115(B9):B09,301. https://doi.org/10.1029/2009JB006914

Akuhara T, Mochizuki K (2015) Hydrous state of the subducting philippine sea plate inferred from receiver function image using onshore and offshore data. J Geophys Res Solid Earth 120(12):8461–8477. https://doi.org/10.1002/2015JB012336

Akuhara T, Mochizuki K, Nakahigashi K, Yamada T, Shinohara M, Sakai S, Kanazawa T, Uehira K, Shimizu H (2013) Segmentation of the Vp/Vs ratio and low-frequency earthquake distribution around the fault boundary of the Tonankai and Nankai earthquakes. Geophys Res Lett 40(7):1306–1310. https://doi.org/10.1002/grl.50223

Akuhara T, Mochizuki K, Kawakatsu H, Takeuchi N (2016) Non-linear waveform analysis for water-layer response and its application to high-frequency receiver function analysis using OBS array. Geophys J Int 206(3):1914–1920. https://doi.org/10.1093/gji/ggw253

Chevrot S (2002) Optimal measurement of relative and absolute delay times by simulated annealing. Geophys J Int 151(1):164–171, https://doi.org/10.1046/j.1365-246X.2002.01755.x

Iritani R, Takeuchi N, Kawakatsu H (2010) Seismic attenuation structure of the top half of the inner core beneath the northeastern Pacific. Geophys Res Lett 37(19):L19,303. https://doi.org/10.1029/2010GL044053

Iritani R, Takeuchi N, Kawakatsu H (2014) Intricate heterogeneous structures of the top 300 km of the Earth's inner core inferred from global array data: I regional 1D attenuation and velocity profiles. Phys Earth Planet Inter 230:15–27. https://doi.org/10.1016/j.pepi.2014.02.002

Kennett BLN, Engdahl ER (1991) Traveltimes for global earthquake location and phase identification. Geophys J Int 105(2):429–465. https://doi.org/10.1111/j.1365-246X.1991.tb06724.x

Kirkpatrick S, Gelatt CD, Vecchi MP (1983) Optimization by Simulated Annealing. Science 220(4598):671–680. https://doi.org/10.1126/science.220.4598.671

Tonegawa T, Iritani R, Kawakatsu H (2013) Extraction of Moho-Generated phases from vertical and radial receiver functions of a seismic array. Bull Seismol Soc Am 103(3):2011–2024. https://doi.org/10.1785/0120120295

Chapter 4
Receiver Function Image of the Subducting Philippine Sea Plate

Abstract We construct receiver function (RF) image using onshore and offshore observatories located around the Kii Peninsula, southwest Japan, to investigate fluid distribution along the subducting Philippine Sea Plate. We calculate RFs at off-shore sites using the inverse water-layer filter method. Resultant RF amplitudes suggest the existence of low-velocity zones directly beneath the plate interface of both onshore and offshore regions. We interpreted this as evidence of hydrous oceanic crust, extending from 5–35 km depth to the plate interface. We attribute the reduction of RF amplitudes beneath the Kii Peninsula to the dehydration of the oceanic crust, which increases the seismic velocity of the oceanic crust. Furthermore, we suggest that the densification caused by the dehydration makes the plate interface permeable. This permeable plate interface may indicate the location of non-volcanic tremors, somewhat contrasting with the situation of long-term slow slip events, which are considered to occur along the impermeable plate interface.

Keywords Receiver function · Southwest Japan subduction zone
Fluid distribution · Dehydration process

4.1 Introduction

Subducting plates undergo metamorphic dehydration reactions due to the surrounding high pressure and temperature at depth. Lines of evidence have suggested that the fluid released by the dehydration reaction affects the slip behavior of megathrust faults. In particular, slow earthquakes, such as non-volcanic tremors and slow slip events (SSEs), are often associated with high pore fluid pressure, which reduces normal effective stress on the plate interface (e.g., Saffer and Tobin 2011). In our study area, around the Kii Peninsula, the subducting Philippine Sea Plate exhibits a complex geometry (e.g., Shiomi et al. 2008). In addition, spatial distribution of slow earthquakes shows differences in the along-strike direction: There is a large gap in a belt-like tremor region where a long-term SSE occured instead of tremors (see Fig. 1.4). Therefore, along-strike variations in fluid distribution are highly expected.

© Springer Nature Singapore Pte Ltd. 2018
T. Akuhara, *Fluid Distribution Along the Nankai-Trough Megathrust Fault off the Kii Peninsula*, Springer Theses, https://doi.org/10.1007/978-981-10-8174-3_4

This chapter aims to constrain the hydrous state of the subducting Philippine Sea Plate mainly from receiver function (RF) amplitudes. We calculate RFs using both on-land stations and ocean-bottom seismometers (OBSs), where the IWLF method described in Chaps. 2 and 3 was applied to the OBS data. The calculated time series of the RFs are then converted in the depth domain to obtain 3-D subsurface structure image. From this image, we construct a 3-D geometry model of the subducting plate. Finally, the RF amplitudes are examined to identify the hydrous state of the Philippine Sea Plate and discuss its relationship with slow earthquakes and intraslab earthquakes.

4.2 Methodology

4.2.1 Common-Conversion-Point Stacking

This section describes the method for constructing RF images of subsurface structures. We first calculated radial RFs for all stations, including the on-land stations (Fig. 1.4), using the ETMTC method with band-pass filters of 0.1–1.0 Hz. For OBS data, we applied inverse water-layer filters (IWLFs) to the vertical-component records prior to the RF calculations, for which we discarded event records showing positive $\Delta_{RMS-ACF}$ values (Sect. 3.3). For the IWLF parameter, we used station-specific values that were estimated in Sect. 3.2 (i.e., the stars in Fig. 3.5a).

The amplitudes of the resulting radial RFs at the OBSs were significantly larger (factor of 3–8) than those at the on-land stations. We believe that this amplification originates from the unconsolidated sediment layer beneath the seafloor and the removal from the OBS vertical-component records of downward P-wave reflection at the seafloor. Note that the latter effect produces relative amplification by a factor of $1 + R$, which is insufficient to explain the observed amplification. The large amplification by the soft sediment seems reasonable considering its extremely high Vp/Vs ratio: ∼2 at 1 km depth from the seafloor, increasing to ∼13 at the seafloor (Hamilton 1979). A more significantly reduced S-wave velocity (compared with P-wave velocity) would lead to more amplified horizontal ground motion than vertical motion.

In Fig. 4.1, we show teleseismic waveforms recorded at different sites and observe significant variation in amplitude among sites even before the application of the IWLF. We attribute this difference to the difference in site responses. To correct this amplification, we calculated the averaged energy ratio of radial-component event records to vertical-component records for each station (Fig. 4.2) and normalized all radial RFs using these ratios (empirical correction of RF amplitudes). The averaged energy ratio was calculated as follows:

Fig. 4.1 Radial (red)- and vertical (black)-component records of teleseismic waveforms at several stations. Vertical bars represent theoretical P-wave arrival times. Light green shaded areas show the time windows used to calculate the average energy ratio of the radial- to vertical-component records. Instrumental responses are removed for all waveforms, but water reverberations are not. Station names are given in the upper right corner of each panel, and the station locations are indicated in Fig. 4.2

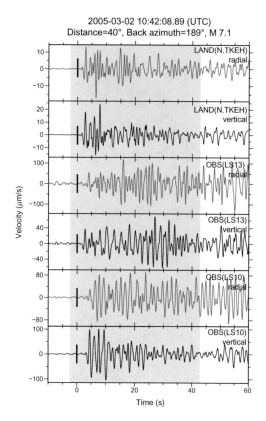

$$Energy\ ratio = \sum_{j} \sqrt{\frac{\int \left\{ u_r^{(j)}(t) \right\}^2 dt}{\int \left\{ u_v^{(j)}(t) \right\}^2 dt}}, \qquad (4.1)$$

where $u_r^{(j)}(t)$ and $u_v^{(j)}(t)$ represent the radial- and vertical-component records of the jth event, respectively.

We then applied time-to-depth conversions to these corrected RFs by referring to 1-D station-specific velocity structures extracted from the 3-D tomography model (Akuhara et al. 2013). As the tomography model employed station correction values that compensated for arrival time delays due to thin sediment layers beneath the seafloor, we shifted the time series of RFs in accordance with the correction values. We applied a uniform shift throughout the entire time series. The averaged station correction values at offshore stations were ~0.0 and ~-0.8 s for P- and S-wave arrivals, respectively (Fig. 4.3a, b). The negative correction value means that observed arrival times were delayed with respect to the synthetic arrival times. After

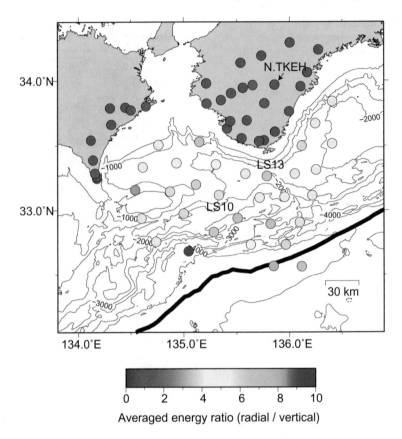

Fig. 4.2 Averaged energy ratio of radial- to vertical-component records for each station (circles). Note that the energy ratio shown here is normalized so that the averaged value of on-land stations is equal to unity

the application of the time shift, we could directly compare the RF image with the tomographic model.

We finally performed common-conversion-point stacking (Dueker and Sheehan 1997) along trough-parallel and trough-normal profiles (X- and Y-axes, respectively) with a separation distance of 10 km (Fig. 4.4a). Each profile contains grid points spaced at 2.5 km horizontally and 0.25 km vertically. First, ray paths for all event records were calculated based on station-specific 1-D velocity structures, event back azimuths, and ray parameters. These calculated ray paths were then discretized vertically with 0.25 km spacing and projected onto profiles within a 10 km distance. Finally, the RF amplitudes were stacked at the grid point closest to the projected points. The same amount was also added to the neighboring four grid points to obtain a smoothed RF image. This process is shown schematically in Fig. 4.4b.

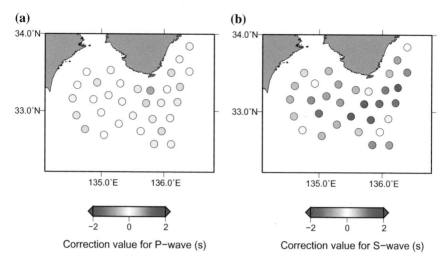

Fig. 4.3 Station correction values for **a** P- and **b** S-wave arrival times (Akuhara et al. 2013)

Fig. 4.4 a Location map of cross sections for common-conversion-point (CCP) stacking. Gray dashed and brown thick lines represent cross sections used in the CCP stacking. On-land stations and ocean-bottom seismometers are shown with the same notation as in Fig. 1.4. Modified from Akuhara and Mochizuki (2015) by permission of John Wiley & Sons Ltd. **b** Schematic illustration of CCP stacking. Note vertical exaggeration. Ray path segment highlighted with pink is projected to five adjoining grids highlighted with green

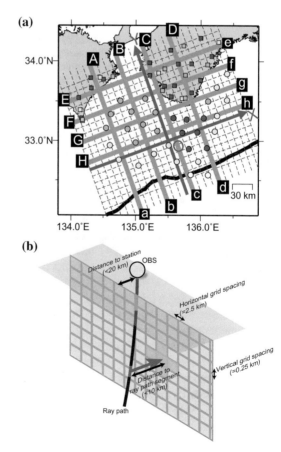

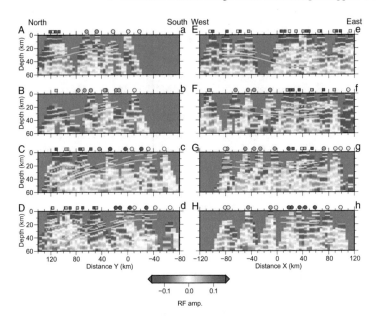

Fig. 4.5 Receiver function image constructed by common-conversion-point stacking. Green solid, dashed, and dotted curves depict the 3-D geometry models of the oceanic Moho, the plate interface, and the island-arc Moho, respectively. Locations of these cross sections are shown in Fig. 4.4. Modified from Akuhara and Mochizuki (2015) by permission of John Wiley & Sons Ltd.

4.3   Results

4.3.1   3-D Geometry of the Subducting Philippine Sea Plate

Figures 4.5 and 4.6 show the common-conversion-point stacking profiles and the P-wave tomography model, respectively. From these figures, we find dominant positive RF amplitudes just above the high-velocity zone of the tomography model (Vp > 7.5 km/s), which can be traced as landward descending curves; we interpret these as the subducting oceanic Moho. To construct the 3-D geometry model of the oceanic Moho, we measured the depths of these RF peaks at as many horizontal grid points as possible and fitted a minimum-curvature surface with tension (Smith and Wessel 1990) to the measured depths (Fig. 4.7a).

The model obtained in this way resolves the geometry of the subducting oceanic Moho from 20–35 km depth in the western region (X < 0 km) and from 15–50 km depth in the eastern region (X > 0 km). Because of poor velocity constraints in the tomography model, we did not include marginal regions in our model; the modeled region is outlined by the light blue line in Fig. 4.7. The modeled Moho forms a depression at Y = 15–55 km along the A–a profile. We consider that this portion represents a subducting seamount identified by a previous active-source survey (Kodaira et al.

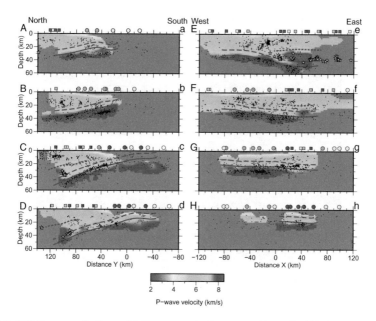

Fig. 4.6 3-D P-wave velocity model from a previous tomographic study (Akuhara et al. 2013). The region where the velocity is not well constrained is masked by a gray color. Blue solid, dashed, and dotted curves depict the 3-D geometry model of the oceanic Moho, the plate interface, and the island-arc Moho, respectively. Green stars and black circles represent low-frequency earthquakes and regular earthquakes relocated by Akuhara et al. (2013), respectively, whose locations are within 10 km of the profiles. The locations of these cross sections are shown in Fig. 4.4. Modified from Akuhara and Mochizuki (2015) by permission of John Wiley & Sons Ltd.

2000, 2002). Our model mostly agrees with the results of previous active-source seismic surveys in offshore regions (Kodaira et al. 2002, 2006; Mochizuki et al. 1998) and RF analyses of onshore regions (Kato et al. 2014; Shiomi et al. 2008).

Ide et al. (2010) proposed the existence of a tear in the subducting PHS Plate along the eastern edge of the subducted Kinan Seamount Chain (almost coinciding with the X = 0 km line) based on the on-land RFs, seismicity, and focal mechanisms of local earthquakes. Our Moho model shows considerable depth difference between the both sides of X = 0 km along the E–e profile (Fig. 4.5). However, from this figure, it is difficult to determine whether this difference in the Moho depth truly reflects the existence of the tear, especially considering the smoothing effect imposed on our model.

Negative amplitudes of RFs can be traced subparallel to, and above, the oceanic Moho, suggesting the presence of a low-velocity zone (LVZ). We can trace this area of negative RF amplitudes in the shallow portion, even beneath the offshore region. We interpret the negative amplitudes as the top of the oceanic crust (i.e., the plate interface), in accordance with previous studies (Bostock 2013, and references therein). Referring to these negative RFs, we estimated the thickness of the oceanic crust to be 8 km. This thickness is consistent with the results of previous active-

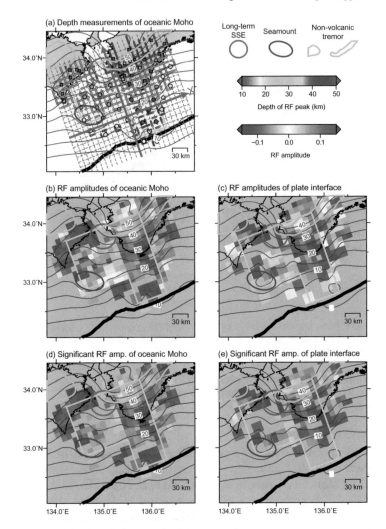

Fig. 4.7 **a** The modeled geometry of the subducting oceanic Moho (brown contours). Small diamonds denote the depth of the receiver function (RF) peak used in the construction of the oceanic Moho model. Circles and squares represent ocean-bottom seismometer and on-land stations, respectively, with the same colors as used in Fig. 1.4. Gray dashed lines show the locations of the vertical profiles for common-conversion-point stacking. Light blue lines enclose the area where the tomography model is well constrained. RF amplitudes of **b** the oceanic Moho and **c** the plate interface. Brown contours in **b** and **c** denote the depths of the oceanic Moho and plate interface, respectively. Brown, pink, and light green lines depict the subducted seamount (Kodaira et al. 2000), slow slip region (Kobayashi 2014), and the belt zone of non-volcanic tremors, respectively. **d, e** The same as **b, c**, respectively, but the area for which the absolute RF amplitude is less than 2σ error is masked. Modified from Akuhara and Mochizuki (2015) by permission of John Wiley & Sons Ltd.

source seismic surveys (Kodaira et al. 2006, 2002; Mochizuki et al. 1998), except near the seamount. In addition, the interpreted plate interface coincides with the focal depths of low-frequency earthquakes (Fig. 4.6, C–c and D–d profiles), which is consistent with the widely accepted notion that low-frequency earthquakes occur on plate interfaces (e.g., Shelly et al. 2006).

4.3.2 Island-Arc Moho Beneath the Kii Peninsula

Beneath the Kii Peninsula, we see continuous positive RF amplitudes above the subducting plate interface at 20–30 km depth (dotted curves in Figs. 4.5 and 4.6). We interpret this plane as the island-arc Moho. Note that the location of intersection between the island-arc Moho and the subducting plate was not resolved in our results. The island-arc Moho shows a northward-dipping feature, which is consistent with the interpretation of Kato et al. (2014).

Below the island-arc Moho interface, the tomography model shows higher velocity (>8.0 km/s) to the north of the tremor (or low-frequency earthquake) zone and lower velocity (~7 km/s) to the south (Fig. 4.6, D–d profile). Although this high-velocity zone is located on the northern edge of the resolvable area, a similar high-velocity zone was seen in another tomography model (Matsubara et al. 2008) that used different datasets with sufficient station coverage to the north (Fig. 4.8). We, therefore, conclude that the lateral velocity change below the island-arc Moho (i.e., the velocity of the topmost mantle) is a robust feature. We attribute this velocity change to serpentinization of the mantle, as discussed in Sect. 4.4.3.

4.3.3 Receiver Function Amplitudes Along the Subducting Plate

We mapped RF amplitudes along the oceanic Moho and plate interface as follows. Initially, we evaluated the RF amplitudes of both the oceanic Moho and the plate interface by taking an average value along the ray path segment penetrating a ±1 km thick volume around the interface. Then, these ray-specific amplitudes were stacked within each half-overlapped 20 × 20 km square bin across our study area. The averaged amplitudes and uncertainties were evaluated for each bin. We employed the bootstrap method with 1000 resampling times to estimate the uncertainties.

Figure 4.7b, c shows the resultant RF amplitudes along the oceanic Moho and the plate interface, respectively. Figure 4.7d, e shows the same amplitudes but the area whose absolute RF amplitude is less than 2σ standard errors is masked. In these figures, we show the amplitudes outside of the modeled region for reference. The oceanic Moho is characterized by mostly positive RF amplitudes. We can trace the negative amplitudes of the plate interface extending seaward to ~5 km depth south

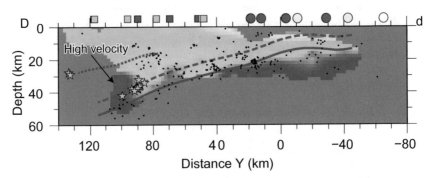

(a) Akuhara et al. [2013]

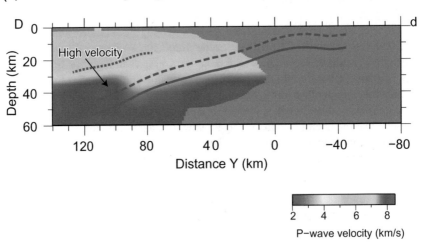

(b) Matsubara et al. [2008]

Fig. 4.8 Comparison of two tomography models by **a** Akuhara et al. (2013) and **b** Matsubara et al. (2008). The locations of these cross sections are exactly identical (D–d profile in Fig. 4.4a). The island-arc Moho, the oceanic Moho, and the plate interface modeled in this study are shown with the same notation as in Fig. 4.6 on both panels

off the Kii Peninsula, which implies that the LVZ exists directly beneath the plate interface.

We can also recognize that the absolute RF amplitudes of both the oceanic Moho and the plate interface are reduced where the top of the PHS slab subducts to ≥35 km depth beneath the Kii Peninsula, whereas its updip portion is characterized by stronger amplitudes. The spatial extent of the zone of reduced amplitudes coincides with that of deep non-volcanic tremors.

4.4 Discussion

4.4.1 Forward Modeling of Amplitude Reduction Beneath the Kii Peninsula

We conducted additional tests to confirm whether the reduced amplitudes in the tremor zone beneath the Kii Peninsula truly reflect the elastic properties. Our concern is that the reduced RF amplitudes may originate from the change in the dip angle of the slab and thus may not represent the elastic properties. The teleseismic events we used are mostly located southward (Fig. 1.5), so their incident angles to the subducting interfaces become more perpendicular as the subduction angle becomes steeper beneath the Kii Peninsula. This could be a cause of the RF amplitude reduction beneath the Kii Peninsula because the efficiency of the P-to-S conversion decreases as the incident angle becomes smaller.

To minimize the effect from the dipping layer, we gathered RFs whose back azimuth was within a $\pm 25°$ range from the perpendicular direction to the dip direction of the slab and within epicentral distances between 35–55°. Then, we stacked these selected RFs for each station and aligned them in order of depth to the oceanic Moho (Fig. 4.9a). The RFs used here were calculated in the same manner as introduced in the Sect. 4.2, but a slightly different frequency bandwidth (0.0–1.0 Hz) was used to suppress the ringing feature.

We manually selected the P-to-S conversion phases from the plate interface and the oceanic Moho on the resultant RF traces (Fig. 4.9a) and measured their amplitudes (Fig. 4.9b). The uncertainties of these amplitudes were estimated from the variations in pre-stack RF amplitudes. We discarded results from the stations whose absolute amplitude was less than 1σ standard error at either the oceanic Moho or the plate interface (triangles in Fig. 4.9b). In addition, we also discarded results from the two most inland stations (squares in Fig. 4.9b) because the assumption of a negative phase existing at the plate interface may not be correct at depth due to the eclogitization of the oceanic crust (Kato et al. 2014). Even after minimizing the slab dip effects in this way, we still see the amplitude reduction with subduction depth (see fitted lines in Fig. 4.9b). Therefore, we conclude that this amplitude reduction reflects the change in the elastic property contrast.

Next, we evaluated the S-wave velocity of the oceanic crust from the observed RF amplitudes along the plate interface. For this, we calculated synthetic RFs using a velocity model composed of an overriding plate with velocity gradient along the depth, a homogeneous oceanic crust, and an oceanic mantle (Fig. 4.10c). The method used to calculate the synthetic RFs was the same as that used in Sect. 2.3, and the ray parameter was set to be the average of the observed data. For each station, we adjusted the S-wave velocity of the oceanic crust so that the synthetic amplitudes of the P-to-S conversion phases from the plate interface could reproduce the fitted lines of the observed RF amplitudes. The P-wave velocity of the oceanic crust and the P- and S-wave velocities of the other layers were derived from the tomography model

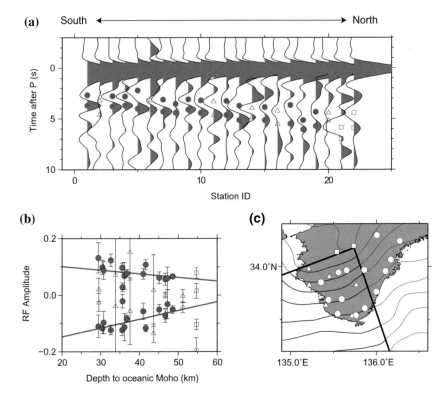

Fig. 4.9 a Radial receiver functions (RFs) stacked at each station on the Kii Peninsula. Positive polarity is shaded gray. Blue and red circles denote manual selections of P-to-S conversion phases from the plate interface and the oceanic Moho, respectively. Triangles and squares represent the same selections but with large uncertainties on their amplitudes and with less confidence in their phase identification, respectively. **b** Relationships between RF amplitudes of the P-to-S conversion phases and subduction depth. Error bars represent the 1σ confidence regions of standard errors. Blue and red lines denote the fitted line to good quality data (circles). **c** Locations of stations. Contour lines represent the oceanic Moho. Thick black lines enclose the area where the tomography model is well constrained

(Akuhara et al. 2013). Density was assumed according to an empirical relationship with P-wave velocity (Birch 1961).

As a result, we found that a drastic increase in the S-wave velocity (from 2.9 to 4.2 km/s) with subduction depth was essential to explain the observed amplitude reduction (Fig. 4.10a, b). According to a phase diagram of MORB (mid-ocean ridge basalt) (Peacock 1993) and a pressure–temperature path simulated for the subducting PHS plate beneath the Kii Peninsula (Yamasaki and Seno 2003), the oceanic crust at the shallow portion (~20 km depth) is considered to correspond to greenschist facies, whereas the deep portion (>40 km depth) is considered to correspond to eclogite facies (Fig. 4.11). However, compared with laboratory experiments, the S-wave velocity of 2.9 km/s is about 25% lower than the velocity expected for green-

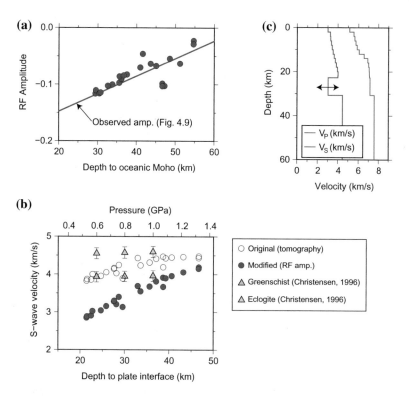

Fig. 4.10 Simple modeling of S-wave velocity of the oceanic crust beneath the Kii Peninsula using RF amplitudes. **a** Blue circles represent synthetic RF amplitudes of the P-to-S conversion phases from the plate interface. Blue line roughly indicates the observed amplitude, as in Fig. 4.9b. **b** Blue circles represent S-wave velocities of the oceanic crust which were modified to explain observed RF amplitudes; open circles show those of the tomography model (Akuhara et al. 2013) for comparison. Light green and light blue triangles denote previous laboratory measurements of S-wave velocities for greenschist and eclogite, respectively (Christensen 1996). Pressure on the upper horizontal axis was calculated for the depth of the plate interface by assuming density to be $2.8 \, g/cm^3$. **c** Typical velocity structure used in the synthetic calculations. We modified only S-wave velocity of the oceanic crust to fit synthetic RF amplitudes to the observed amplitudes

schist facies (Fig. 4.10b) (Christensen 1996). We suggest that the oceanic crust therein hosts abundant fluid, thereby producing such a low velocity at shallow depth. The highest velocity, 4.2 km/s, is close to the shear velocity of eclogite facies, but is still 10% lower than laboratory measurements (Fig. 4.10b). We also attribute this 10% reduction in S-wave velocity to fluid content, but note that the amount of fluid would be smaller than that of the updip portion.

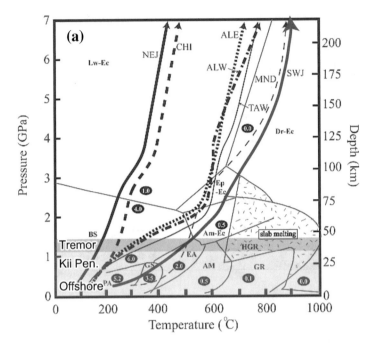

Fig. 4.11 Pressure–temperature (P–T) path along the subducting Philippine Sea Plate beneath the Kii Peninsula (red curve) and the phase diagram for the water-saturated MORB (modified after Yamasaki and Seno 2003). The P–T path was calculated by (Yamasaki and Seno, 2003) and passes pumpellyite-actinolite (PA), greenschist (GS), epidote-amphibolite (EA), and amphibole-eclogite (Am-Ec) facies. The shallow region of the phase diagram, below the eclogite facies, is based on Peacock (1993). Note that the phase diagram presented here omits some phases which appear at low temperature (e.g., zeolite facies). Outlined numbers represent the maximum amount of H_2O (wt%). Explanations of the other notations can be seen in the caption of Fig. 4a of Yamasaki and Seno (2003). Shaded areas with pale blue, yellow, and light green highlight the depth ranges of the subducting plate interface inferred from the present study beneath the offshore region, the Kii Peninsula, and the tremor belt, respectively. Modified from Yamasaki and Seno (2003) by permission of John Wiley and Sons Ltd.

4.4.2   Uncertainty of Relative Depth Between Hypocenters And the Oceanic Moho

In Fig. 4.12, we show the hypocenters relocated by Akuhara et al. (2013) and the geometry of the PHS slab obtained in this study. Both the slab model and the seismicity are based on the same tomography model (Akuhara et al. 2013), so a direct comparison of their depths is possible. In this section, we aim to evaluate the uncertainties in the relative depths of these local earthquakes to the oceanic Moho. We derive a rough estimate of the uncertainties using typical values. In the following discussion, we adopt 6.8 and 3.8 km/s as typical values of P- and S-wave velocities, respectively, which represent the velocities of the oceanic crust. In addition, we use

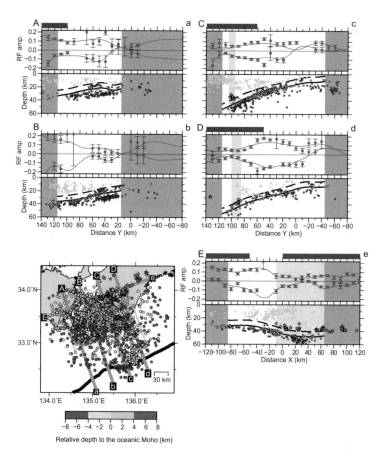

Fig. 4.12 Distribution of intraslab earthquakes, low-frequency earthquakes (LFEs), and receiver function (RF) amplitudes on the plate interface and oceanic Moho. Blue and red circles denote intraslab earthquakes occurring within the oceanic crust and the oceanic mantle, respectively. Those earthquakes within 4 km distance from the oceanic Moho are discriminated using a pale color. Light green stars represent LFEs. In the lower panel of each cross section, dashed and solid black curves indicate the plate interface and oceanic Moho modeled in this study, respectively. Earthquakes occurring above the plate interface are indicated by gray circles. All hypocenters are located within 10 km of the profiles. In the upper panel of each cross section, blue and red circles show RF amplitudes on the plate interface and oceanic Moho along the profiles, respectively, while blue and red curves show the laterally smoothed versions. Thick brown bars appearing above the upper panels show the extent of land area. Gray and green shaded areas correspond to the regions of the unconstrained tomography model and the range of LFE occurrences, respectively. Modified from Akuhara and Mochizuki (2015) by permission of John Wiley & Sons Ltd.

0.55 Hz, corresponding to the central frequency of our band-pass filter, to estimate the typical wavelengths of the RFs.

We initially evaluated the uncertainties in the depth estimates of the oceanic Moho. We considered that these uncertainties originate from two sources: (1) a lack of a high-frequency component in RFs due to band-pass filtering and (2) the assumptions of 1-D velocity models without any dipping structures. We assume that the first type of uncertainty, referred to as σ_{m1}, is equal to one-eighth of a typical RF wavelength, that is, $\sigma_{m1} \sim 2$ km. The second type of uncertainty (σ_{m2}) has been well discussed by Shiomi et al. (2004), who investigated how depth estimations of a dipping oceanic Moho are affected by the assumption of a horizontal oceanic Moho. In accordance with their work, we assume $\sigma_{m2} \sim 3$ km. This value was calculated for an oceanic Moho with a 15° dip angle at 30 km depth.

For hypocenter errors (σ_h), we utilized root mean squared residual values between observed and synthetic travel times from a previous tomography analysis (Akuhara et al. 2013). On average, the values are 0.08 s and 0.12 s for P- and S-waves, respectively. Using typical values for oceanic crustal velocity, we estimated the uncertainty of hypocenters to be less than 1 km.

Combining the three types of error, σ_{m1}, σ_{m2}, and σ_h, into a single composite uncertainty, $\sqrt{\sigma_{m1}^2 + \sigma_{m2}^2 + \sigma_h^2}$, we assumed that the relative location error of hypocenters to the oceanic Moho would be ~ 4 km at most. Other than these three kinds of uncertainty, the uncertainties in the 1-D velocity models themselves and the station correction values should be taken into consideration for error estimations of the absolute depths. However, we were able to neglect such factors when discussing the relative location errors because the errors for the hypocenters and the oceanic Moho would be mutually canceled out.

4.4.3   Hydrous State of the Subducting Oceanic Crust

To date, many studies have used RFs to infer the hydrous state of subducting slabs in many subduction zones (Bostock 2013, and references therein). Generally, based on observations of dominant negative RF amplitudes along a plate interface overlying positive RF amplitudes of the oceanic Moho, these studies support the idea that oceanic crust hosts abundant fluid. The most likely origin of the fluid is dehydration of the subducting slabs, which has been supported by numerous studies (e.g., Hacker et al. 2003).

In accordance with these studies, we interpret that negative RF amplitudes along the plate interface ranging from 5–35 km depth (Fig. 4.7e) reflect the existence of fluid below the interface. Although this interpretation lacks quantitative verification, it is in good agreement with other independent observations in this region by Kodaira et al. (2002) and Kurashimo et al. (2013). They performed active-source seismic surveys in offshore and onshore regions of this study area and identified the reflective phase characteristics of trapped fluid along the plate interface. The possible sources of the fluid are a series of metamorphic dehydration reactions of the oceanic crust, including zeolite, pumpellyite-actinolite, greenschist, epidote-amphibolite and amphibole-eclogite (Fig. 4.11), and porosity collapse of the oceanic crust (Hyndman and Peacock 2003). An important implication of this interpretation is that the plate interface must be impermeable, or sealed, to trap fluid within the oceanic crust. This permeability barrier might originate from small-grained silicic mudrock, foliations of phyllosilicates formed by coseismic rupture, and hydrothermal precipitation of silicic materials (Sibson 2013, and references therein).

In the following, we infer the hydrous state of the subducting PHS plate. We divided our study area into three regions: the Kii Peninsula, the southeastern Shikoku Island, and the offshore region.

The Kii Peninsula

We first focus on the Kii Peninsula, where the PHS plate subducts to 20–45 km depth and RF amplitudes are reduced significantly with the slab subduction. Given the pressure–temperature path reported by Yamasaki and Seno (2003) and the phase diagram of MORB from Peacock (1993), metamorphic phase change from greenschist to epidote-amphibolite occurs within the oceanic crust (Fig. 4.11). As we have already stated in Sect. 4.4.1, the RF amplitude reduction can be explained by a combination of the velocity increase due to the metamorphic phase change and the loss of fluid content within the oceanic crust. We interpret that the loss of fluid (notwithstanding the fluid production during the metamorphic phase change) is caused by increasing permeability near the plate interface. At the deeper portion, a successive dehydration reaction toward eclogite would be accompanied by densification. Such densification may produce fracture within the oceanic crust, which results in the increase in permeability and allows fluid to ascend into the overriding plate (Fig. 4.13a). A similar interpretation was made for Cascadia subduction zone (Bostock 2013).

There are two independent observations that may be applicable to our interpretation. The first observation is the velocity of the overriding plate, as predicted by the tomography model (Akuhara et al. 2013; Matsubara et al. 2008). From Fig. 4.8, we see that a relatively low P-wave velocity (~7 km/s) is located within the island-arc mantle just above the subducting slab top to 30–40 km depth. The velocity is too low for typical rocks of the mantle. In accordance with laboratory experiments (Christensen 2004), we interpret this low-velocity portion as serpentinized mantle (i.e., hydrated mantle rock). We consider that ascending fluid from the oceanic crust passing through the permeable plate interface contributes to serpentinization (Fig. 4.13a). Further inland beyond the tremor zone, the velocity of the island-arc mantle becomes higher (>8 km/s). We interpret this as non-serpentinized mantle. We postulate that the dehydration reaction of the oceanic crust has been completed at depth; therefore, no fluid would ascend into the island-arc mantle.

The second observation is the earthquake distribution determined by Akuhara et al. (2013). Along the dip direction beneath the Kii Peninsula (along the C–c and D–d profiles in Fig. 4.12), we see that the RF amplitudes tend to be strong in the seismogenic portion of the oceanic crust. One possibility for this is that these earthquakes within the oceanic crust are promoted by elevated pore fluid pressure resulting from metamorphic phase change from greenschist to epidote-amphibolite facies (Fig. 4.11). In fact, a rapid rate of fluid production has been reported during this phase change (Kuwatani et al. 2011). The downdip offset of the seismogenic crust may coincide with the termination of this phase change. The aseismic oceanic crust at

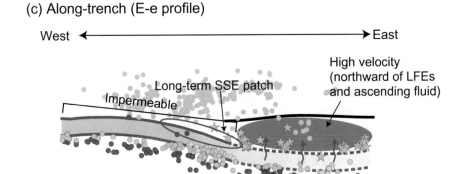

(a) Kii Peninsula (C-c profile)

North ← — — — — — — — — → South

High velocity   Serpentinized   Impermeable

Ascending fluid
Dehydration completed

(b) Southeastern Shikoku (A-a profile)

North ← — — — — — — → South

Impermeable   Seamount

(c) Along-trench (E-e profile)

West ← — — — — — — — — — → East

High velocity
(northward of LFEs
and ascending fluid)

Long-term SSE patch
Impermeable

Fig. 4.13 Schematic illustration of our interpretation along the profiles in Fig. 4.12: **a** C–c, **b** A–a, and **c** E–e profiles. Circles and stars denote earthquakes and low-frequency earthquakes with the same color usage as that in Fig. 4.12. Blue, red, and black curves show the plate interface, oceanic Moho, and island-arc Moho, respectively. Modified from Akuhara and Mochizuki (2015) by permission of John Wiley & Sons Ltd.

further downdip is considered to undergo the next step of the dehydration reaction, from epidote-amphibolite to eclogite. Abers et al. (2013) have proposed that the eclogitization of this region is accompanied by a high rate of solid volume reduction that prevails over the fluid volume expansion rate. This is a suitable condition for prohibiting seismicity and developing fractures that increase the permeability.

Previous studies also reported the existence of LVZs or high-reflectivity zones along the subducting plate beneath the Kii Peninsula and associated these with the slab dehydration reaction (Kato et al. 2014; Kurashimo et al. 2013). The benefit of the present study over previous studies is that our discussion is based on a quantified analysis using RF amplitudes. With this feature, we could interpret along-dip changes in fluid content and the permeability of the plate interface. The other advantage is that we conducted 3-D analysis. The resultant map view of the RF amplitudes (Fig. 4.7) allows us to recognize a good spatial correlation between the reduced RF amplitudes and the source region of non-volcanic tremors. This is further discussed in Sect. 4.4.4.

Southeastern Shikoku Island

Beneath the southeastern edge of Shikoku Island, such a significant RF amplitude reduction as beneath the Kii Peninsula cannot be seen. Because the subduction depth is shallow beneath this region, dehydration of the oceanic crust would not proceed such as it did at the deeper portion beneath the Kii Peninsula. We therefore conclude that the plate interface is sufficiently impermeable to strongly retain the RF amplitudes (Fig. 4.13b). The zone of minor seismicity within the oceanic crust (Fig. 4.12, $X = 50$–70 km of the A–a profile) is characterized by strong RF amplitudes. If associating the strong amplitude with the high-velocity contrast on the basis of the high degree of fluid content, one may consider that this seismicity is enhanced by high pore fluid pressure. However, relatively large errors in the RF amplitudes make it difficult to reach any confident conclusions in this regard.

Offshore Region

Our results show that the plate interface beneath the offshore region is characterized by negative RF amplitudes, suggesting the existence of the LVZ below the interface. Since the seismic velocity of typical dry oceanic crust is higher than that of the basement of the accretionary prism (i.e., the overriding plate), we consider that fluid contributes to the LVZs. While the presence of such LVZs below (or perhaps along) the plate interface has been identified by previous active-source surveys (Kodaira et al. 2002; Kamei et al. 2012; Bangs et al. 2009), our study is the first to reveal the LVZ extends over >100 km in width in the trench-parallel direction.

Throughout this chapter, we take the RF amplitudes along the plate interface and the oceanic Moho as indicators of velocity contrast at the corresponding interface. This assumption is plausible for on-land stations, but perhaps not for OBSs. This is because the sediment layer beneath the seafloor produces strong S-wave reverberations within the layer. Since we used a low-frequency content (<1.0 Hz) to image the subsurface structure, the amplitudes of P-to-S conversions from the plate interface (or the oceanic Moho) may be contaminated by such reverberations. We conduct high-frequency analysis in Chap. 5 to isolate such reverberation effects.

4.4.4 Permeability Difference in Source Area of Long-Term Slow Slip Events and Tremors

Our results demonstrate the spatial correlation between tremors and the reduced RF amplitudes, and we have interpreted that the plate interface is permeable due to fractures induced by eclogitization. This interpretation is somewhat in contrast to the relationship between the impermeable plate interface and long-term SSEs, which has been proposed by previous studies (Kato et al. 2010; Kodaira et al. 2004; Song et al. 2009). We speculate that such differences in permeability could be a factor distinguishing long-term SSE and tremor zones. That is, an impermeable plate interface effectively traps fluid to reduce the effective normal stress over a wide area of the plate interface. In contrast, a fractured plate interface can accumulate pore fluid pressure only within small patches defined by the typical scale of the fracture.

Although we have little constraint on the RF amplitudes of the 1996–1997 long-term SSE region at the Kii Channel, the region is most likely to be imaged as intense RF amplitudes based on the amplitudes of the neighboring region at the same depth range (25–30 km depth to the plate interface, see Fig. 4.7d and e). We therefore consider it likely that the plate interface at the long-term SSE region is impermeable, or at least less permeable than that of the tremor zone (Fig. 4.13c).

References

Abers GA, Nakajima J, van Keken PE, Kita S, Hacker BR (2013) Thermal-petrological controls on the location of earthquakes within subducting plates. Earth Planet Sci Lett 369–370:178–187. https://doi.org/10.1016/j.epsl.2013.03.022

Akuhara T, Mochizuki K (2015) Hydrous state of the subducting philippine sea plate inferred from receiver function image using onshore and offshore data. J Geophys Res Solid Earth 120(12):8461–8477. https://doi.org/10.1002/2015JB012336

Akuhara T, Mochizuki K, Nakahigashi K, Yamada T, Shinohara M, Sakai S, Kanazawa T, Uehira K, Shimizu H (2013) Segmentation of the Vp/Vs ratio and low-frequency earthquake distribution around the fault boundary of the Tonankai and Nankai earthquakes. Geophys Res Lett 40(7):1306–1310. https://doi.org/10.1002/grl.50223

Bangs N, Moore G, Gulick S, Pangborn E, Tobin H, Kuramoto S, Taira A (2009) Broad, weak regions of the Nankai Megathrust and implications for shallow coseismic slip. Earth Planet Sci Lett 284(1–2):44–49. https://doi.org/10.1016/j.epsl.2009.04.026

Birch F (1961) The velocity of compressional waves in rocks to 10 kilobars: 2. J Geophys Res 66(7):2199–2224. https://doi.org/10.1029/JZ066i007p02199

Bostock MG (2013) The Moho in subduction zones. Tectonophysics 609:547–557. https://doi.org/10.1016/j.tecto.2012.07.007

Christensen NI (1996) Poisson's ratio and crustal seismology. J Geophys Res 101(B2):3139–3156. https://doi.org/10.1029/95JB03446

Christensen NI (2004) Serpentinites, peridotites, and seismology. Int Geol Rev 46(9):795–816. https://doi.org/10.2747/0020-6814.46.9.795

Dueker KG, Sheehan AF (1997) Mantle discontinuity structure from midpoint stacks of converted P to S waves across the Yellowstone hotspot track. J Geophys Res 102(B4):8313–8327. https://doi.org/10.1029/96JB03857

Hacker BR, Peacock SM, Abers GA, Holloway SD (2003) Subduction factory 2. Are intermediate-depth earthquakes in subducting slabs linked to metamorphic dehydration reactions? J Geophys Res 108(B1):2030. https://doi.org/10.1029/2001JB001129

Hamilton EL (1979) Vp/Vs and Poisson's ratios in marine sediments and rocks. J Acoust Soc Am 66(4):1093–1101. https://doi.org/10.1121/1.383344

Hyndman RD, Peacock SM (2003) Serpentinization of the forearc mantle. Earth Planet Sci Lett 212(3-4):417–432. https://doi.org/10.1016/S0012-821X(03)00263-2

Ide S, Shiomi K, Mochizuki K, Tonegawa T, Kimura G (2010) Split Philippine Sea plate beneath Japan. Geophys Res Lett 37(21):L21, 304. https://doi.org/10.1029/2010GL044585

Kamei R, Pratt RG, Tsuji T (2012) Waveform tomography imaging of a megasplay fault system in the seismogenic Nankai subduction zone. Earth Planet Sci Lett 317–318:343–353. https://doi.org/10.1016/j.epsl.2011.10.042

Kato A, Iidaka T, Ikuta R, Yoshida Y, Katsumata K, Iwasaki T, Sakai S, Thurber C, Tsumura N, Yamaoka K, Watanabe T, Kunitomo T, Yamazaki F, Okubo M, Suzuki S, Hirata N (2010) Variations of fluid pressure within the subducting oceanic crust and slow earthquakes. Geophys Res Lett 37(14):L14, 310. https://doi.org/10.1029/2010GL043723

Kato A, Saiga A, Takeda T, Iwasaki T, Matsuzawa T (2014) Non-volcanic seismic swarm and fluid transportation driven by subduction of the Philippine Sea slab beneath the Kii Peninsula, Japan. Earth Planets Space 66(1):86. https://doi.org/10.1186/1880-5981-66-86

Kobayashi A (2014) A long-term slow slip event from 1996 to 1997 in the Kii Channel, Japan. Earth Planets Space 66(1):9. https://doi.org/10.1186/1880-5981-66-9

Kodaira S, Takahashi N, Nakanishi A, Miura S, Kaneda Y (2000) Subducted seamount imaged in the rupture zone of the 1946 Nankaido earthquake. Science 289(5476):104–106. https://doi.org/10.1126/science.289.5476.104

Kodaira S, Kurashimo E, Park JO, Takahashi N, Nakanishi A, Miura S, Iwasaki T, Hirata N, Ito K, Kaneda Y (2002) Structural factors controlling the rupture process of a megathrust earthquake at the Nankai trough seismogenic zone. Geophys J Int 149(3):815–835. https://doi.org/10.1046/j.1365-246X.2002.01691.x

Kodaira S, Iidaka T, Kato A, Park JO, Iwasaki T, Kaneda Y (2004) High pore fluid pressure may cause silent slip in the Nankai trough. Science 304(5675):1295–1298. https://doi.org/10.1126/science.1096535

Kodaira S, Hori T, Ito A, Miura S, Fujie G, Park JO, Baba T, Sakaguchi H, Kaneda Y (2006) A cause of rupture segmentation and synchronization in the Nankai trough revealed by seismic imaging and numerical simulation. J Geophys Res 111(B9):B09, 301. https://doi.org/10.1029/2005JB004030

Kurashimo E, Iwasaki T, Iidaka T, Kato A, Yamazaki F, Miyashita K, Shibutani T, Ito K, Takeda T, Obara K, Hirata N (2013) Along-strike structural changes controlled by dehydration-related fluids within the Philippine Sea plate around the segment boundary of a megathrust earthquake beneath the Kii peninsula, southwest Japan. Geophys Res Lett 40(18):4839–4844. https://doi.org/10.1002/grl.50939

Kuwatani T, Okamoto A, Toriumi M (2011) Thermodynamic forward modeling of progressive dehydration reactions during subduction of oceanic crust under greenschist facies conditions. Earth Planet Sci Lett 307(1–2):9–18. https://doi.org/10.1016/j.epsl.2011.01.027

Matsubara M, Obara K, Kasahara K (2008) Three-dimensional P- and S-wave velocity structures beneath the Japan Islands obtained by high-density seismic stations by seismic tomography. Tectonophysics 454(1–4):86–103. https://doi.org/10.1016/j.tecto.2008.04.016

Mochizuki K, Fujie G, Sato T, Kasahara J, Hino R, Shinohara M, Suyehiro K (1998) Heterogeneous crustal structure across a seismic block boundary along the Nankai Trough. Geophys Res Lett 25(13):2301–2304. https://doi.org/10.1029/98GL51867

Peacock SM (1993) The importance of blueschist eclogite dehydration reactions in subducting oceanic crust. Geol Soc Am Bull 105(5):684–694. https://doi.org/10.1130/0016-7606(1993)105<0684:TIOBED>2.3.CO;2

Saffer DM, Tobin HJ (2011) Hydrogeology and mechanics of subduction zone forearcs: fluid flow and pore pressure. Ann Rev Earth Planet Sci 39(1):157–186. https://doi.org/10.1146/annurev-earth-040610-133408

Shelly DR, Beroza GC, Ide S, Nakamula S (2006) Low-frequency earthquakes in Shikoku, Japan, and their relationship to episodic tremor and slip. Nature 442(7099):188–191. https://doi.org/10.1038/nature04931

Shiomi K, Sato H, Obara K, Ohtake M (2004) Configuration of subducting Philippine Sea plate beneath southwest Japan revealed from receiver function analysis based on the multivariate autoregressive model. J Geophys Res 109(B4):B04, 308. https://doi.org/10.1029/2003JB002774

Shiomi K, Matsubara M, Ito Y, Obara K (2008) Simple relationship between seismic activity along Philippine Sea slab and geometry of oceanic Moho beneath southwest Japan. Geophys J Int 173(3):1018–1029. https://doi.org/10.1111/j.1365-246X.2008.03786.x

Sibson RH (2013) Stress switching in subduction forearcs: implications for overpressure containment and strength cycling on megathrusts. Tectonophysics 600:142–152. https://doi.org/10.1016/j.tecto.2013.02.035

Smith WHF, Wessel P (1990) Gridding with continuous curvature splines in tension. Geophysics 55(3):293–305. https://doi.org/10.1190/1.1442837

Song TRA, Helmberger DV, Brudzinski MR, Clayton RW, Davis P, Perez-Campos X, Singh SK (2009) Subducting slab ultra-slow velocity layer coincident with silent earthquakes in southern Mexico. Science 330(5926):502–506. https://doi.org/10.1126/science.1167595

Yamasaki T, Seno T (2003) Double seismic zone and dehydration embrittlement of the subducting slab. J Geophys Res 108(B4):2212. https://doi.org/10.1029/2002JB001918

Chapter 5
A Fluid-Rich Layer Along the Megathrust Fault Inferred from High-Frequency Receiver Function Inversion Analysis

Abstract In this chapter, we conduct receiver function inversion analysis to investigate detailed seismic properties near the megathrust fault using ocean-bottom seismometers deployed off the Kii Peninsula, southwest Japan. RFs were calculated at high frequencies (up to 4 Hz), removing the effect of water reverberations from the vertical-component records. Our inversion was performed in two steps: first, we modeled the sediment layer by a simple stacking method; second, we determined deeper structure using a waveform inversion. The results indicate the presence of a thin low-velocity zone (LVZ) of a 0.2–1.2 km thickness with an S-wave velocity of 0.7–2.4 km/s along the plate interface. We interpret this LVZ as a thin fluid-rich sediment layer between the overriding and subducting plates, which acts as a pathway for fluid migration.

Keywords Receiver function · Ocean-bottom seismometer
Southwest Japan subduction zone · Waveform inversion · Fluid-rich layer

5.1 Introduction

In the previous chapter, we found that the subducting Philippine Sea Plate interface is characterized by negative receiver function (RF) amplitudes even beneath offshore region. These negative amplitudes can be considered evidence for the existence of a low-velocity zone (LVZ) directly beneath the plate interface at the seismogenic zone. A possible explanation for this LVZ is the hydrated oceanic crust, as discussed in the previous chapter. This interpretation partly originates from the thickness of the LVZ ($\sim$8 km, see Fig. 4.5). However, the thickness of the LVZ tends to be overestimated if the imaging process (i.e., time–depth conversion) is based on tomography models, because the tomographic method tends to underestimate velocity anomalies owing to its smoothing constraint (Song and Helmberger 2007).

Also in the last chapter, we quantitatively evaluated the S-wave velocity of the LVZ beneath the land area of the Kii Peninsula based on RF amplitudes (4.4.1).

© Springer Nature Singapore Pte Ltd. 2018
T. Akuhara, *Fluid Distribution Along the Nankai-Trough Megathrust Fault off the Kii Peninsula*, Springer Theses, https://doi.org/10.1007/978-981-10-8174-3_5

Such quantitative analysis is useful for interpreting the physical properties of the LVZ. However, in the case of ocean-bottom seismometers (OBSs) our empirical correction to the sediment amplification effects may not be sufficient to deduce the physical properties of the LVZ from the amplitudes of P-to-S conversions.

Taking into account the above points, it is desirable to quantitatively evaluate the physical properties and thickness of the LVZ by extracting as much information as possible from the RF waveforms. In doing so for the case of OBS data, we must pay careful attention to sediment reverberations. We must also conduct high-frequency analysis to isolate such reverberations. In this chapter, we first assess the physical properties of the sediment layer by the H-κ stacking method (Zhu and Kanamori 2000). Then, we perform RF inversion analysis to investigate the deeper subsurface structure, especially the LVZ.

5.2 High-Frequency Receiver Functions

We calculated the radial RFs of the OBSs (Fig. 5.1), employing the ETMTC and the IWLF method[1] with a low-pass filter at 4.0 Hz. Such high-frequency contents were necessary to distinguish a number of sediment reverberations. The event records employed here are in the same as those used in Chap. 4, but we discarded the data from deep earthquakes occurring around Japan so as to suppress variations in ray parameters among events sharing similar back azimuths.

Although the resultant radial RFs show coherent signals over the traces at each station, they are difficult to interpret at first sight. This complexity is most likely due to unconsolidated sediment covering the seafloor. Out of all 32 OBSs, we selected five OBSs, LS02, LS03, LS05, LS09, and LS11, whose RFs exhibited clear positive peaks 1–2 s after direct P arrival (Fig. 5.2a–e; a typical OBS not showing such peaks are shown in Fig. 5.2f for reference). We interpret this positive peak as a P-to-S conversion from the bottom of the sediment layer. Another common feature among the five OBSs is the absence of positive peaks at zero-lag time. The low seismic velocity of the sediment causes near-vertical incidence of the direct P phase, leading to no energy on the horizontal components. Several coherent phases can be seen after the arrival of the P-to-S conversions, which may be interpreted as sediment-related reverberations or P-to-S conversions from deeper interfaces. We find successive negative and positive phases (blue and red arrows in Fig. 5.2a–e) near the arrival times of the P-to-S conversions from the subducting plate interface that is expected from

---

[1] We used slightly different values for the WLF parameters, τ and R from those estimated in Chap. 3; The values used in this section were determined by a minorly modified method (Akuhara et al. 2016) from that of Chap. 3, where we did not solve for the normalization factors (see also a footnote in Sect. 3.2.1). We confirmed that this minor change in τ and R did not affect our conclusion.

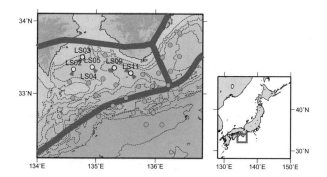

Fig. 5.1 Tectonic setting of the study area and locations of ocean-bottom seismometers (OBSs) (yellow and gray circles). Blue lines enclose the source regions of the 1944 Tonankai (to the east) and 1946 Nankai (to the west) earthquakes. The five OBSs featured in this study are highlighted in yellow. Pink sectors show the back azimuthal bins of the teleseismic events to be analyzed. The location of the study area is enclosed by a red rectangle in the right-hand panel. Reproduced from Akuhara et al. (2017) by permission of John Wiley and Sons Ltd.

the established tomography model (Akuhara et al. 2013). The short time intervals between these phases (<1 s) suggest a thin LVZ along the subducting plate interface. However, we cannot and should not judge whether this prediction is correct or not before identifying multiple phases from the sediment layer (Kawakatsu and Abe 2016). Hereinafter, these negative and positive phases are referred to as PsL− and PsL+ phases, respectively.

In the subsequent sections, we first estimate the sediment layer properties using the H-κ stacking method (Zhu and Kanamori 2000) to confirm that the PsL− and PsL+ phases originate not from the sediment layer beneath the seafloor, but from the LVZ along the subducting plate interface (Sect. 5.3). We then conduct RF waveform inversion to investigate the LVZ properties (Sect. 5.4). We set a limited back azimuth bin to be analyzed for each of LS02, LS03, LS05, and LS09, and two back azimuth bins for LS11 (hereafter, we refer to these two bins as LS11N and LS11S and use these terms as station names). We located these back azimuth bins where the PsL− and PsL+ phases have relatively strong amplitudes (Fig. 5.2). Throughout this paper, we consider only a 1-D layered isotropic structure, and second-order features such as a dipping interface and anisotropy will be neglected for simplicity, although we find that the PsL− and PsL+ phases show a move-out pattern that is typical of a landward dipping interface at some stations (e.g., the PsL− and PsL+ phases arrive earlier for LS11S than they do for LS11N).

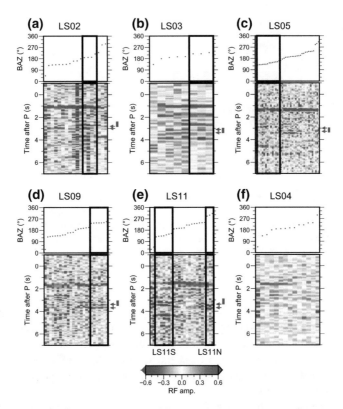

Fig. 5.2 Radial receiver functions (RFs) (bottom row) and their event back azimuths (top row) for **a–e** five OBSs with clear P-to-S conversions from the bottom of the sediment and **f** a typical OBS that does not show such clear conversions. Thick black lines enclose the RFs analyzed in this study. Red and blue arrows denote the PsL+ and PsL− phases, respectively. Blue bars represent the depths of the subducting plate interface estimated in Chap. 4, where the time–depth conversion was conducted using a previous tomography model (Akuhara et al. 2013). Reproduced from Akuhara et al. (2017) by permission of John Wiley and Sons Ltd.

5.3   Estimation of Sediment Properties by the H-κ Stacking Method

5.3.1   Methodology

We applied the H-κ stacking analysis (Zhu and Kanamori 2000) to investigate the properties of the sediment layer beneath the seafloor. In this method, thickness, h, and Vp/Vs ratio, κ, of a homogeneous layer are determined via a grid search so that the theoretical timings of the conversion and reflection phases can yield constructive interference. In addition to three commonly used phases, Ps, PpPs, and PpSs, we used two additional phases, PsSs and PpPs+w, which tend to be dominant on OBS

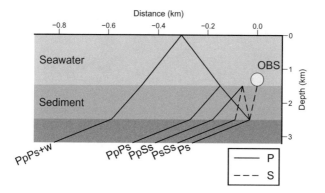

Fig. 5.3 Definition of phase names used in the H-κ stacking method. The ray paths were calculated using typical sediment velocities (1.7 km/s and 0.43 km/s for P- and S-wave velocities, respectively) and a ray parameter of 0.07 s·km^{-1}. Note horizontal exaggeration. Modified from Akuhara et al. (2017) by permission of John Wiley and Sons Ltd.

records (see Fig. 5.3 for the definition of these phase names). The delay times of these phases from a direct P-wave can be expressed as follows:

$$\Delta T_{Ps} = h \left(\sqrt{\left(\frac{\kappa}{V_P}\right)^2 - p^2} - \sqrt{\frac{1}{V_P^2} - p^2} \right) \tag{5.1}$$

$$\Delta T_{PpPs} = h \left(\sqrt{\left(\frac{\kappa}{V_P}\right)^2 - p^2} + \sqrt{\frac{1}{V_P^2} - p^2} \right) \tag{5.2}$$

$$\Delta T_{PpSs} = 2h \left(\sqrt{\left(\frac{\kappa}{V_P}\right)^2 - p^2} \right) \tag{5.3}$$

$$\Delta T_{PsSs} = h \left(3\sqrt{\left(\frac{\kappa}{V_P}\right)^2 - p^2} - \sqrt{\frac{1}{V_P^2} - p^2} \right) \tag{5.4}$$

$$\Delta T_{PpPs+w} = h \left(\sqrt{\left(\frac{\kappa}{V_P}\right)^2 - p^2} + \sqrt{\frac{1}{V_P^2} - p^2} \right) + \tau. \tag{5.5}$$

In these equations, V_P, p, and τ represent the P-wave velocity of the sediment layer, the ray parameter, and the two-way travel time within the water layer, respectively. We assumed V_P to be 1.7 km/s (Hamilton 1979), so h and κ were unknown parameters.

Using Eqs. 5.1–5.5, we performed a grid search on the h-κ space to find the maximum of the stack function:

$$s = \sum_{i=1}^{N} \left\{ w_1 r_i(\Delta T_{Ps}) + w_2 r_i(\Delta T_{PpPs}) - w_3 r_i(\Delta T_{PpSs}) - w_4 r_i(\Delta T_{PsSs}) + w_5 r_i(\Delta T_{PpPs+w}) \right\},$$

(5.6)

where $r_i(t)$ represents a receiver function of the ith trace out of N traces and $w_1, w_2, \ldots, w_5$ represent weighting factors assigned for each phase type. In this study, we set $w_1 = 0.5$, $w_2 = w_3 = 0.05$, and $w_4 = w_5 = 0.2$ so that the phases with larger expected amplitudes have a more significant influence on the estimation.

5.3.2   Results

Figure 5.4 shows the results of the H-κ stacking analysis for each OBS. Relatively large stacked amplitudes define a hyperbola-like curve in the h-κ space, which is dominated by Ps phases. Two different data sets from a single OBS (LS11S and LS11N) produce similar results, indicating the stability of our analysis. The estimated thickness and Vp/Vs ratio from the maximum stacked amplitudes vary considerably among OBSs: the estimated thickness ranges from 0.6 to 1.1 km, and the Vp/Vs ratio from 3.0 to 5.3. Such variations may be reasonable considering that the separation distance of the OBSs (20 km) is larger than or comparable with the typical length scale in the undulation of the sediment thickness in this region ($\sim$10 km) (Tsuji et al. 2015). A previous reflection survey identified clear reflectors at the bottom of the fore-arc basin sediment in this region, and the reflectors were characterized by a 0.5– 2.0 s two-way travel time from the seafloor (Tsuji et al. 2015). This two-way travel time corresponds to a 0.4–1.7 km thickness if we assume its P-wave velocity to be 1.7 km/s; this is roughly consistent with the thickness we estimated. The Vp/Vs ratio of the sediment layer may also vary in response thickness variation (Hamilton 1979). Unfortunately, no study has reported the Vp/Vs ratio of the sediment layer in our study area. A Vp/Vs ratio of 2.85–3.67 has been estimated off Shikoku Island, but this is 200–300 km away from our study area (Takahashi et al. 2002); our estimated values are comparable, or somewhat higher, than this estimation.

Based on the above result of the H-κ stacking, we then computed synthetic RFs using a three-layer model composed of water, sediment, and half-space layers to compare them with the observed RFs. Velocities of the half-space layer were given by the tomography model (Akuhara et al. 2013), and thickness and velocities of the sediment layer were given by the results of the H-κ stacking analysis. The synthetic RFs were produced by first calculating both radial and vertical synthetic waveforms using the propagator matrix method (Haskell 1953). Then, we computed radial RFs from these two components using IWLF and a low-pass filter at 4.0 Hz. Observed RFs were stacked in the frequency domain using the frequency-dependent uncertainties as weighting factors (Park and Levin 2000). Standard errors of the stacked RFs were estimated by bootstrap sampling 2000 times (Efron 1982).

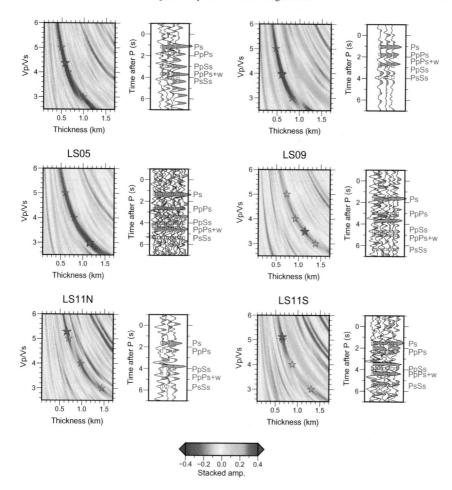

Fig. 5.4 Results of *H*-κ stacking analysis. (Left column) The stacked amplitudes in *H*-κ spaces. Red solid and open stars on the panels indicate the best sediment model predicted by the maximum amplitudes and additional models with fixed V_P/V_S values, respectively. (Right column) The receiver functions used in the analysis. Red and blue squares denote the predicted arrival of sediment-related phases by the best model with positive and negative polarities, respectively. Reproduced from Akuhara et al. (2017) by permission of John Wiley and Sons Ltd.

The resultant synthetic RFs well reproduced the observed amplitudes of the direct P-wave arrivals, Ps phases, and PpPs phases for most stations, whereas the PsL– and PsL+ phases were not recovered by this experiment (Fig. 5.5, red solid traces). Thus, we interpret that the PsL– and PsL+ phases are P-to-S conversions originating from the top and bottom of the thin LVZ along the subducting plate interface. Since the PsL– and PsL+ phases can be seen for all OBSs deployed over a zone ∼100 km wide, we postulate that the LVZ is a dominant structure of the subducting plate interface, rather than local heterogeneity. Considering the dominant hyperbola-like

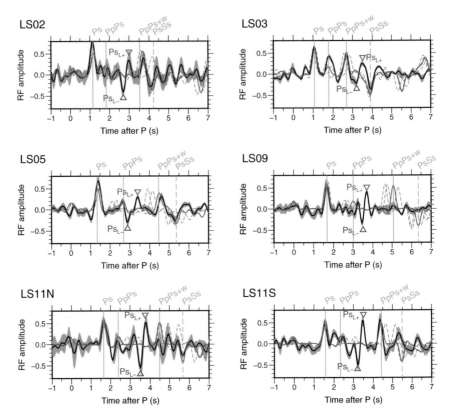

Fig. 5.5 Comparison of observed and synthetic receiver functions (RFs). Synthetic RFs (red traces) were computed using a three-layered model composed of water, sediment, and a half-space layer, where the sediment layer is based on the best results of H-κ stacking analysis. Stacked observed RFs are shown as black traces with 2σ standard errors shown by gray shading. Green solid and dashed lines represent the predicted arrivals of positive and negative phases, respectively. Blue and red triangles denote major non-modeled phases that we interpret as P-to-S conversion phases from a low-velocity zone along the subducting plate interface, i.e., PsL– and PsL+ phases. Modified from Akuhara et al. (2017) by permission of John Wiley and Sons Ltd.

curves of the Ps phases in Fig. 5.4, our H-κ stacking analysis might suffer from a trade-off between thickness and Vp/Vs ratio. Therefore, we tested three additional sediment models with Vp/Vs ratios of 3.0, 4.0, and 5.0 (Fig. 5.4, red open stars), in addition to the model predicted by the maximum stacked amplitudes (Fig. 5.4, red solid stars). The results show that all of three sediment models fail to reproduce both PsL– and PsL+ phases at all stations (Fig. 5.5, red dashed traces), further supporting our interpretation.

5.4 Receiver Function Inversion Analysis

5.4.1 Methodology

Our inversion searches for 1-D isotropic layered models that well predict the observed RFs. We assume that the 1-D layered structure is composed of six homogeneous layers including the bottom half-space: i.e., from top to bottom, the seawater, sediment, overriding plate, LVZ along the plate interface, oceanic crust, and oceanic mantle, (Fig. 5.6). To fully describe the model, we need P- and S-wave velocities (V_P and V_S, respectively), density (ρ), and thickness (h) for each layer. Inverting all parameters without any prior knowledge, however, seems unrealistic, considering the non-uniqueness of RF inversion (Ammon et al. 1990). Instead, we limited the number of model parameters and search ranges by exploiting prior information gained from the results of the H-κ stacking analysis and other previous studies as follows. Note that we imposed relatively loose restrictions on the LVZ parameters as they were the main targets.

For the water layer, the P-wave velocity and density were fixed at typical values ($V_P = 1.5\,\text{km/s}$ and $\rho = 1.0\,\text{g/cm}^3$). The thickness was also fixed at the OBS depths, which had been measured by acoustic ranging. No unknown parameters were assigned to this layer. For the sediment layer, all parameters (h, V_P, V_S, and ρ) were fixed and the values were exported from the best model of the H-κ stacking analysis. Like the water layer, no unknown parameters were assigned to this layer.

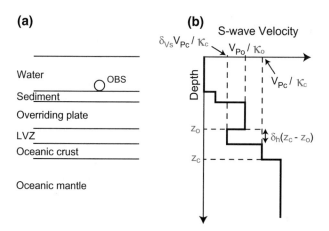

Fig. 5.6 Schematic illustration of layered structure with depth-velocity diagrams, as assumed in our inversion. The six parameters shown in red (z_o, z_c, κ_o, κ_c, δ_h, and δ_{V_S}) are model parameters. The other parameters were fixed using prior information. P-wave velocities of the overriding plate and the oceanic crust are denoted by V_{P_o} and V_{P_c}, respectively. Note that the S-wave velocity and thickness of the low-velocity zone are expressed by $\delta_{V_S} V_{P_c}/\kappa_c$ and $\delta_h(z_c z_o)$, respectively. Reproduced from Akuhara et al. (2017) by permission of John Wiley and Sons Ltd.

For the overriding and oceanic crust layers, we treated the layer bottom depths (z_o and z_c, respectively) and Vp/Vs ratio (κ_o and κ_c, respectively) as unknown parameters (Fig. 5.6). The search range of zo and zc were set to ± 2 km wide around the depth of the plate interface and the oceanic Moho, respectively, which is estimated in Chap. 4. The search ranges of κ_o and κ_c were 1.6–1.8 and 1.6–2.0, respectively. We extracted the P-wave velocity from the tomography model (Akuhara et al. 2013). In addition, we specified density using an empirical relationship with P-wave velocity (Birch 1961) as follows:

$$\rho = 0.828 V_P + 0.613 \tag{5.7}$$

We parameterized the LVZ layer using two special parameters, δ_h and δ_{V_S}, which represent the percentage thickness and S-wave velocity of this layer relative to those of the oceanic crust layer, respectively. With this treatment, we could allow the LVZ to reside around the plate interface depth. Note that the absolute thickness and S-wave velocities of this layer are given by $\delta_h(z_c - z_o)$ and $\delta_{V_S} V_{P_c}/\kappa_c$, respectively, where V_{P_c} represents the P-wave velocity of the oceanic crust (Fig. 5.6); their search ranges were set at 0–50% for δ_h and 10–80% for δ_{V_S}. The density was specified using the empirical relationship in Eq. 5.7. For the oceanic mantle, all parameters were fixed at typical values ($V_P = 8.1$ km/s, $V_S = 4.7$ km/s, and $\rho = 3.4$ g/cm^3).

To find the optimum parameter set, we employed the neighborhood algorithm, a variant of Monte Carlo simulation (Sambridge 1999). The algorithm first randomly generates 4000 samples in the multidimensional parameter space and evaluates misfit values of these samples. At every iteration step, the algorithm selects the 20 samples with the lowest misfit values out of all samples so far generated. Then, two new samples are randomly generated in the proximity of each selected sample. We repeated this process 400 times to generate 20000 models.

We defined the misfit values based on cross-correlation coefficients as follows (Frederiksen et al. 2003):

$$misfit = 1 - \frac{\int_t u_{syn}(t) \cdot u_{obs}(t)dt}{\sqrt{\int_t u_{syn}^2(t)dt} \cdot \sqrt{\int_t u_{obs}^2(t)dt}} \tag{5.8}$$

where u_{syn} and u_{obs} represent synthetic and observed RFs, respectively. We used 8 s time windows starting 1 s before the direct P-wave arrival to the misfit measurements. This window length was chosen so that the PsSs phase could be included.

5.4.2   Results

Our inversion produced well-converged misfit values after the 400 iteration steps (Fig. 5.7). Considering the uncertainties in the observed RFs, ranges of the parameters that can well explain the observed waveforms are more important than the best fit result. We therefore selected "preferable" models out of all 20000 models using two

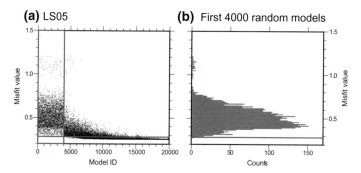

Fig. 5.7 Misfit values of individual models generated during inversion analysis for a typical OBS, LS05. **a** All misfit values obtained during the inversion are shown by dots, aligned in chronological order. The blue rectangle encloses the 4000 models generated randomly at the first iteration step. The red line denotes the threshold value for determining preferable models. **b** Histogram of misfit values for the first 4000 random models. The threshold value (red line) is defined as the 0.1 percentile of this distribution

criteria. As the first criterion, we employed a threshold misfit value for each OBS. We defined the threshold value as the fourth lowest misfit value (i.e., 0.1 percentile) of the 4000 random models generated at the first iteration step of the inversion (Fig. 5.7). In other words, any model with misfit values lower than this threshold is expected to show a better waveform fit than random samples with a probability of 0.999. The second criterion was introduced to ensure good waveform fits for the PsL– and PsL+ phases. For this purpose, we selected models that predicted both PsL– and PsL+ phases within ±0.15 s ranges around the peaks of the observed PsL– and PsL+ phases. Note that the 0.15 s range corresponds to a quarter of the dominant period of the observed RFs.

Figure 5.8 shows the ensembles of S-wave velocity profiles of the extracted preferable models. As we expected, the velocity profiles show thin LVZs near the plate interface depths. Their velocities are significantly lower than the upper limit we imposed for the inversion (i.e., 80% of the oceanic crust velocity). The velocity profiles of LS11N and LS11S are consistent, except at the LVZ depth. This difference most likely represents the landward dipping interface as seen from the move-out patterns of the PsL– and PsL+ phases (Fig. 5.2e). We also see that the synthetic waveforms of the preferable models well reproduce the observed PsL– and PsL+ phases (Fig. 5.9). The synthetic seismograms calculated for pure elastic (i.e., no attenuation) structures tended to overestimate the amplitudes of some later phases (e.g., PpPs+w for LS09 and LS11N, and PsSs for LS09 and LS11S). We attribute this partly to topography variations of the sediment–basement interface in the fore-arc accretionary prism (Tsuji et al. 2015) (particularly severe for PpPs+w, which has a long horizontal leg (Fig. 5.3)), and partly to the strong attenuation in the sediment layer (Hino et al. 2015).

The preferable models indicate a positive correlation, or a trade-off, between S-wave velocity and thickness of the LVZs; this is reasonably well mitigated with

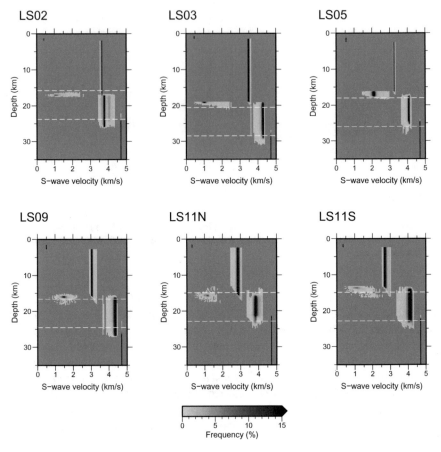

Fig. 5.8 Ensemble of preferable S-wave velocity structures obtained by inversion analysis (brown color). White dashed lines denote the depths of the plate interface and the oceanic Moho estimated by Akuhara and Mochizuki (2015). The area not sampled by the preferable models is masked by gray. Reproduced from Akuhara et al. (2017) by permission of John Wiley and Sons Ltd.

constraints from the amplitudes of the PsL– and PsL+ phases (Fig. 5.10). Here, we define the preferable ranges of the LVZ parameters so that all preferable models are included in the ranges (red rectangles in Fig. 5.10). The averaged preferable ranges of all stations were 0.7–2.4 km/s for S-wave velocity and 0.2–1.2 km for thickness (Fig. 5.11). We also conducted the waveform inversion and evaluated the preferable parameter ranges using the different sediment models (i.e., red open stars in Fig. 5.4). Consequently, we found that our results do not change markedly (Fig. 5.12). This seems reasonable based on Fig. 5.5, in which the sediment models do not produce sediment-related phases overprinting the PsL– and PsL+ phases.

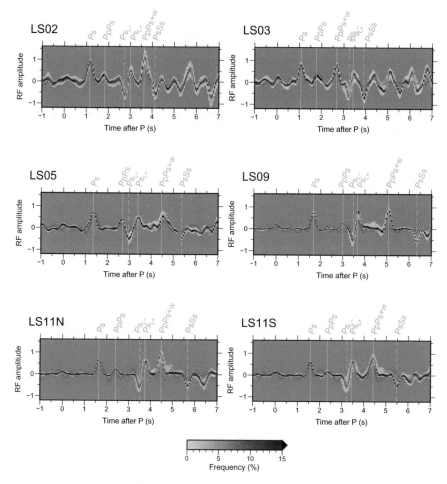

Fig. 5.9 Ensemble of synthetic receiver functions (RFs) calculated with preferable models obtained by inversion analysis (brown color). Red solid and dashed traces show observed RFs and their $2sigma$ standard error, respectively. Green solid and dashed lines represent the predicted arrivals of positive and negative phases, respectively. The area where no synthetic traces pass is masked by gray. Reproduced from Akuhara et al. (2017) by permission of John Wiley and Sons Ltd.

5.5 Fluid-Rich Subducting Sediment Layer Along the Plate Interface

We have characterized an LVZ along the plate interface beneath an offshore region at seismogenic zone depths. The averaged preferable ranges were 0.7–2.4 km/s for S-wave velocity and 0.2–1.2 km for thickness (Fig. 5.11). A number of RF studies (and tomographic analyses) have identified LVZs along subducting oceanic plates in various subduction zones and interpreted them as hydrous oceanic crust (Bostock

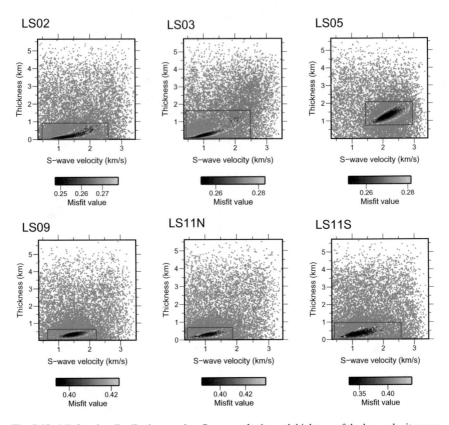

Fig. 5.10 Misfit value distributions against S-wave velocity and thickness of the low-velocity zone (LVZ). A brownish color is given for preferable models, whereas the other models are shown by gray dots. Red lines denote the preferable ranges of the S-wave velocity and thickness of the LVZ. Reproduced from Akuhara et al. (2017) by permission of John Wiley and Sons Ltd.

2013). However, our estimated thickness is too thin to be interpreted as the whole oceanic crust, which should be ∼7 km thick (Kodaira et al. 2006), or the upper oceanic crust, which is ∼3 km thick (Hansen et al. 2012).

We interpret that the LVZ reflects a thin fluid-rich sediment layer between the subducting plate and overriding plate (i.e., along the subducting plate interface). Similar interpretations have been made from active-source seismic surveys conducted in our study area based on intense P-to-P reflection phases from the plate interface (Kodaira et al. 2002). Although the thickness of such a subducting sediment layer has not been estimated at the depth where we identified the LVZ (15–20 km depth), it is estimated to have been 1–2 km before the subduction (Ike et al. 2008) and at shallower subduction depths (<10 km) (Bangs et al. 2009; Kamei et al. 2012; Tsuji et al. 2014). We consider that a similar fluid-rich sediment layer extends beneath our study area, possibly with its thickness tapering off during the subduction.

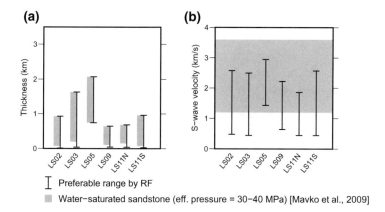

I Preferable range by RF

▨ Water−saturated sandstone (eff. pressure = 30−40 MPa) [Mavko et al., 2009]

Fig. 5.11 Summary of low-velocity zone (LVZ) parameters; **a** thickness and **b** S-wave velocity, as obtained by this study. Vertical bars denote preferable ranges of the parameters for each station. Blue shaded area in Fig. 5.11b represents the S-wave velocity range of sandstone, as reported by laboratory measurements (Mavko et al. 2009), and the area in Fig. 5.11a shows the corresponding thickness range of preferable models. Reproduced from Akuhara et al. (2017) by permission of John Wiley and Sons Ltd.

The obtained S-wave velocities, 0.7–2.4 km/s, are suitable for water-saturated sedimentary rocks; for example, according to laboratory experiments (Mavko et al. 2009 and references therein), water-saturated sandstone under an effective pressure of 30–40 MPa can explain the LVZ velocities (Fig. 5.11). Note that the effective pressure of 30–40 MPa corresponds to ∼10–20% of the vertical load from the overriding plate at the LVZ depth. It also corresponds to an overpressure ratio (Screaton et al. 2002) of ∼0.7–0.9, if simply assuming a lithostatic pressure equal to the vertical load and a pore pressure equal to the difference between the lithostatic and effective pressures. This overpressure ratio is somewhat lower than that reported for the shallower subduction depth where very low-frequency earthquakes occur (Tsuji et al. 2014). Although we only present sandstone as an example here, other types of sedimentary rock (e.g., mudstone) are plausible LVZ materials. Moreover, a lower velocity may be explained by a high degree of clay content (Eberhart-Phillips et al. 1989).

As incoming sediment is considered fully dehydrated at the early stages of subduction (Hyndman and Peacock 2003), we infer that the low-velocity sediment layer is a fluid pathway rather than a fluid source. The existence of a fluid pathway that transports fluid from deep to shallow along the subducting Philippine Sea Plate has been suggested by heat flow modeling (Spinelli and Wang 2008) and by the lithium isotope ratio in submarine mud volcano fluid (Nishio et al. 2015).

Similarly, thin LVZs have been revealed through active-source seismic surveys at seismically locked zones of other subduction zones based on intense P-to-P reflection phases from the plate interface (Nedimović et al. 2003; Mochizuki et al. 2005; Bell et al. 2010; Li et al. 2015), suggesting that subducting fluid-rich sediment layers are a ubiquitous feature of subduction zones. Nedimović et al. (2003) and Li et al. (2015)

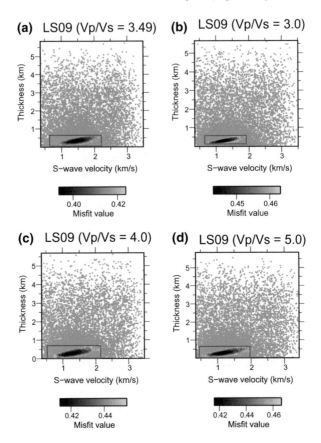

Fig. 5.12 Misfit value distributions against LVZ parameters of a typical station, LS09, derived from different seafloor sediment models: **a** the best model based on *H*-*κ* stacking analysis, (**b**–**d**) alternative models with fixed Vp/Vs ratios of **b** 3.0, **c** 4.0, and **d** 5.0. Notations are the same as in Fig. 5.10

estimated the thicknesses of the LVZs at Cascadia and Alaska subduction zones to be <2 km and 100–250 m, respectively, which is roughly consistent with our results. They also revealed an abrupt change in the thickness along the dip direction: the LVZs become thicker (>2 km) at greater depth.

References

Akuhara T, Mochizuki K (2015) Hydrous state of the subducting philippine sea plate inferred from receiver function image using onshore and offshore data. J Geophys Res Solid Earth 120(12):8461–8477. https://doi.org/10.1002/2015JB012336

Akuhara T, Mochizuki K, Nakahigashi K, Yamada T, Shinohara M, Sakai S, Kanazawa T, Uehira K, Shimizu H (2013) Segmentation of the Vp/Vs ratio and low-frequency earthquake distribution around the fault boundary of the Tonankai and Nankai earthquakes. Geophys Res Lett 40(7):1306–1310. https://doi.org/10.1002/grl.50223

Akuhara T, Mochizuki K, Kawakatsu H, Takeuchi N (2016) Non-linear waveform analysis for water-layer response and its application to high-frequency receiver function analysis using OBS array. Geophys J Int 206(3):1914–1920. https://doi.org/10.1093/gji/ggw253

Akuhara T, Mochizuki K, Kawakatsu H, Takeuchi N (2017) A fluid-rich layer along the Nankai trough megathrust fault off the kii peninsula inferred from receiver function inversion. J Geophys Res Solid Earth 122:6524–6537. https://doi.org/10.1002/2017JB013965

Ammon CJ, Randall GE, Zandt G (1990) On the nonuniqueness of receiver function inversions. J Geophy Res 95(B10):15,303. https://doi.org/10.1029/JB095iB10p15303

Bangs N, Moore G, Gulick S, Pangborn E, Tobin H, Kuramoto S, Taira A (2009) Broad, weak regions of the Nankai Megathrust and implications for shallow coseismic slip. Earth Planet Sci Lett 284(1–2):44–49. https://doi.org/10.1016/j.epsl.2009.04.026

Bell R, Sutherland R, Barker DHN, Henrys S, Bannister S, Wallace L, Beavan J (2010) Seismic reflection character of the Hikurangi subduction interface, New Zealand, in the region of repeated Gisborne slow slip events. Geophys J Int 180(1):34–48. https://doi.org/10.1111/j.1365-246X.2009.04401.x

Birch F (1961) The velocity of compressional waves in rocks to 10 kilobars: 2. J Geophys Res 66(7):2199–2224. https://doi.org/10.1029/JZ066i007p02199

Bostock MG (2013) The Moho in subduction zones. Tectonophysics 609:547–557. https://doi.org/10.1016/j.tecto.2012.07.007

Eberhart-Phillips D, Han DH, Zoback MD (1989) Empirical relationships among seismic velocity, effective pressure, porosity, and clay content in sandstone. Geophysics 54(1):82–89. https://doi.org/10.1190/1.1442580

Efron B (1982) The jackknife, the bootstrap and other resampling plans. Society for Industrial and Applied Mathematics, Philadelphia

Frederiksen AW, Folsom H, Zandt G (2003) Neighbourhood inversion of teleseismic Ps conversions for anisotropy and layer dip. Geophys J Int 155(1):200–212. https://doi.org/10.1046/j.1365-246X.2003.02043.x

Hamilton EL (1979) Vp/Vs and Poisson's ratios in marine sediments and rocks. J Acoust Soc Am 66(4):1093–1101. https://doi.org/10.1121/1.383344

Hansen RT, Bostock MG, Christensen NI (2012) Nature of the low velocity zone in Cascadia from receiver function waveform inversion. Earth Planet Sci Lett 337-338:25–38. https://doi.org/10.1016/j.epsl.2012.05.031

Haskell NA (1953) The dispersion of surface waves on multilayered media. Bull Seismol Soc Am 43(1):17–34

Hino R, Tsuji T, Bangs NL, Sanada Y, Park JO, von Huene R, Moore GF, Araki E, Kinoshita M (2015) QP structure of the accretionary wedge in the Kumano Basin, Nankai Trough, Japan, revealed by long-offset walk-away VSP. Earth Planet Space. https://doi.org/10.1186/s40623-014-0175-x

Hyndman RD, Peacock SM (2003) Serpentinization of the forearc mantle. Earth Planet Sci Lett 212(3-4):417–432. https://doi.org/10.1016/S0012-821X(03)00263-2

Ike T, Moore GF, Kuramoto S, Park JO, Kaneda Y, Taira A (2008) Tectonics and sedimentation around Kashinosaki Knoll: a subducting basement high in the eastern Nankai Trough. Island Arc 17(3):358–375. https://doi.org/10.1111/j.1440-1738.2008.00625.x

Kamei R, Pratt RG, Tsuji T (2012) Waveform tomography imaging of a megasplay fault system in the seismogenic Nankai subduction zone. Earth Planet Sci Lett 317–318:343–353. https://doi.org/10.1016/j.epsl.2011.10.042

Kawakatsu H, Abe Y (2016) Comment on nature of the seismic lithosphere-asthenosphere boundary within normal oceanic mantle from high-resolution receiver functions by Olugboji et al. Geochem Geophys Geosyst 17(8):3488–3492. https://doi.org/10.1002/2016GC006418

Kodaira S, Kurashimo E, Park JO, Takahashi N, Nakanishi A, Miura S, Iwasaki T, Hirata N, Ito K, Kaneda Y (2002) Structural factors controlling the rupture process of a megathrust earthquake at the Nankai Trough seismogenic zone. Geophys J Int 149(3):815–835. https://doi.org/10.1046/j.1365-246X.2002.01691.x

Kodaira S, Hori T, Ito A, Miura S, Fujie G, Park JO, Baba T, Sakaguchi H, Kaneda Y (2006) A cause of rupture segmentation and synchronization in the Nankai Trough revealed by seismic imaging and numerical simulation. J Geophys Res 111(B9):B09,301. https://doi.org/10.1029/2005JB004030

Li J, Shillington DJ, Bécel A, Nedimović MR, Webb SC, Saffer DM, Keranen KM, Kuehn H (2015) Downdip variations in seismic reflection character: implications for fault structure and seismogenic behavior in the Alaska subduction zone. J Geophys Res Solid Earth 120(11):7883–7904. https://doi.org/10.1002/2015JB012338

Mavko G, Mukerji T, Dvorkin J (2009) The rock physics handbook: tools for seismic analysis of porous media. Cambridge University Press, New York

Mochizuki K, Nakamura M, Kasahara J, Hino R, Nihisno M, Kuwano A, Nakamura Y, Yamada T, Shinohara M, Sato T, Moghaddam PP, Kanazawa T (2005) Intense PP reflection beneath the aseismic forearc slope of the Japan Trench subduction zone and its implication of aseismic slip subduction. J Geophys Res 110(B1):B01,302. https://doi.org/10.1029/2003JB002892

Nedimović MR, Hyndman RD, Ramachandran K, Spence GD (2003) Reflection signature of seismic and aseismic slip on the Northern Cascadia subduction interface. Nature 424(6947):416–420. https://doi.org/10.1038/nature01840

Nishio Y, Ijiri A, Toki T, Morono Y, Tanimizu M, Nagaishi K, Inagaki F (2015) Origins of lithium in submarine mud volcano fluid in the Nankai accretionary wedge. Earth Planet Sci Lett 414:144–155. https://doi.org/10.1016/j.epsl.2015.01.018

Park J, Levin V (2000) Receiver functions from multiple-taper spectral correlation estimates. Bull Seismol Soc Am 90(6):1507–1520. https://doi.org/10.1785/0119990122

Sambridge M (1999) Geophysical inversion with a neighbourhood algorithm-I. Searching a parameter space. Geophys J Int 138(2):479–494. https://doi.org/10.1046/j.1365-246X.1999.00876.x

Screaton E, Saffer D, Henry P, Hunze S (2002) Porosity loss within the underthrust sediments of the Nankai accretionary complex: implications for overpressures. Geology 30(1):19–22. https://doi.org/10.1130/0091-7613(2002)030<0019:PLWTUS>2.0.CO;2

Song TRA, Helmberger DV (2007) Validating tomographic model with broad-band waveform modelling: an example from the LA RISTRA transect in the Southwestern United States. Geophys J Int 171(1):244–258. https://doi.org/10.1111/j.1365-246X.2007.03508.x

Spinelli GA, Wang K (2008) Effects of fluid circulation in subducting crust on Nankai margin seismogenic zone temperatures. Geology 36(11):887. https://doi.org/10.1130/G25145A.1

Takahashi N, Kodaira S, Nakanishi A, Park JO, Miura S, Tsuru T, Kaneda Y, Suyehiro K, Kinoshita H, Hirata N, Iwasaki T (2002) Seismic structure of Western end of the Nankai trough seismogenic zone. J Geophys Res Solid Earth 107(B10). https://doi.org/10.1029/2000JB000121

Tsuji T, Kamei R, Pratt RG (2014) Pore pressure distribution of a mega-splay fault system in the Nankai Trough subduction zone: insight into up-dip extent of the seismogenic zone. Earth Planet Sci Lett 396:165–178. https://doi.org/10.1016/j.epsl.2014.04.011

Tsuji T, Ashi J, Strasser M, Kimura G (2015) Identification of the static backstop and its influence on the evolution of the accretionary prism in the Nankai Trough. Earth Planet Sci Lett 431:15–25. https://doi.org/10.1016/j.epsl.2015.09.011

Zhu L, Kanamori H (2000) Moho depth variation in Southern California from teleseismic receiver functions. J Geophys Res 105(B2):2969–2980. https://doi.org/10.1029/1999JB900322

Chapter 6
General Discussion

Abstract In the previous chapters, we developed a new method for conducting receiver function (RF) analysis using ocean-bottom seismometer (OBS) data and estimated the hydrous state of the subducting Philippine Sea Plate. In this chapter, we briefly discuss the following topics, which are not mentioned much in the previous chapters, but are worthy of studying in the future: (1) along-dip variation of LVZ thickness along the subducting plate, (2) a comparison of different subduction zones regarding slow earthquakes, and (3) an expected perspective based on high-frequency RF analysis using OBS data. All of these topics may be addressed in the future by applying our method, which will lead to advanced knowledge relating to the physical properties of megathrust faults.

Keywords Receiver function · Ocean-bottom seismometer · Subduction zones Low-velocity zone

6.1 Along-Dip Variation in Low-Velocity Zone Thickness

We have investigated the hydrous state of the Philippine Sea Plate using two different methods: receiver function (RF) imaging (Chap. 4) and the RF inversion (Chap. 5). The major advantage of the imaging method lies in its capacity to reveal 3-D heterogeneous structures. Through this method, we acquired insight regarding the geometry of the subducting structure and the hydrous state of the PHS plate, which vary in both trough-parallel and trough-normal directions. However, the imaging methods may not effectively constrain the LVZ properties, due to a lack of high-frequency contents, contamination by reverberation phases, and errors in the reference velocity model. In particular, by using smoothed reference velocity models, such as tomography models, the imaging method will underestimate the LVZ thickness (e.g., Hansen et al. 2012).

Our RF imaging identified an ∼8-km thick LVZ extending broadly beneath the onshore and offshore regions. However, this result should be further tested using another method that is more independent of the tomography model. This is because the tomography model we used in this study has smooth velocity changes with depth,

© Springer Nature Singapore Pte Ltd. 2018
T. Akuhara, *Fluid Distribution Along the Nankai-Trough Megathrust Fault off the Kii Peninsula*, Springer Theses, https://doi.org/10.1007/978-981-10-8174-3_6

probably leading to underestimation of the low-velocity anomaly. In addition to our study, a number of studies have conducted RF analysis in southwestern Japan (e.g., Shiomi et al. 2008; Kato et al. 2014), and they also employed smooth velocity models. We therefore consider that the accurate thickness of the LVZ beneath southwestern Japan has not been clearly defined.

In this sense, our RF inversion analysis for some ocean-bottom seismometer (OBS) data (Chap. 5) provides a more accurate assessment of LVZ thickness (<2 km) at the specific portion of the southwestern Japan subduction zone. Note that the thickness of the LVZ is not necessarily constant throughout the entire subduction zone. In fact, Nedimović et al. (2003); Li et al. (2015) reported that the LVZ thickness increases as the slab subducts deeper in the Cascadia and Alaska subduction zones, respectively.

We may recognize such spatial variation of the LVZ thickness in our study area from the time intervals between the PsL− and PsL+ phases. Figure 6.1 shows the radial RFs of two OBSs and two on-land stations. We find that the time intervals between the two phases are longer at on-land stations than at OBSs. It is considered that the LVZ becomes thicker as slab subducts deeper. From Fig. 6.1, we determined that the typical time intervals are ∼0.5 s beneath the offshore region and ∼1.0 s beneath the onshore region. If, for simplicity, we assume a constant seismic velocity and dip angle of the LVZ beneath the four stations, the LVZ is twice as thick beneath on-land stations as it is beneath OBSs.

We may associate this change in LVZ thickness with a fluid process of the subducting PHS plate. As discussed in Sect. 4.4.3, the dehydration reaction from greenschist to epidote-amphibolite facies takes place beneath the Kii Peninsula (Fig. 4.11). Therefore, the thickened LVZ beneath the Kii Peninsula is considered to represent the hydrated oceanic crust (Fig. 6.2). In contrast, beneath the offshore region where the PHS plate subducts to <20 km depth, the pressure–temperature condition of the subducting crust is not suitable for major dehydration reactions to occur. As discussed in Sect. 5.5, we consider that the thin LVZ identified using OBS data represents an incoming sediment layer that acts as a path way for fluid. Moreover, the oceanic crust beneath the sediment layer may not be as hydrated as the downdip portion beneath the Kii Peninsula (Fig. 6.2).

Nedimović et al. (2003); Li et al. (2015) reported similar changes in LVZ thickness at the Cascadia and Alaska subduction zones, respectively. They argue that seismogenic zones (i.e., where megathrust earthquakes occur) are characterized by a thin LVZ, whereas transition zones (i.e., the downdip of the seismogenic zone) are characterized by a thick LVZ. Whether the same is true for our study area is unclear because of (i) the limited constraint on the rupture zones of past megathrust earthquakes and (ii) the sparseness of stations to which we applied RF inversion analysis. Further analysis using a greater number of OBSs and on-land stations may help address this issue.

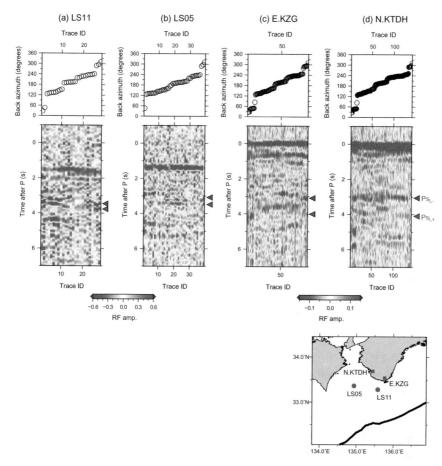

Fig. 6.1 Radial receiver functions (RFs) of **a, b** ocean-bottom seismometers (OBSs) and **c, d** on-land stations plotted along their back azimuths. Red and blue triangles indicate PsL- and PsL+ phases, respectively). Note that time interval between PsL- and PsL+ phases tends to be longer at on-land stations than OBSs. The locations of stations are shown in the bottom inset

6.2 Comparison to Other Subduction Zones

Slow earthquakes, such as tremors and slow slip events (SSEs), are commonly observed in many subduction zones of young age, but their patterns of occurrence are slightly different. For example, in southwestern Japan, tremors define a belt-shaped source region of ∼20 km width in the along-dip direction. However, in the Cascadia subduction zone, tremors are distributed more widely in the along-dip direction (∼100 km). Moreover, the Cascadia subduction zone has no 'long-term' SSE occurrence unlike southwestern Japan. To comprehend such a difference, a comparison of the seismic velocities of the LVZs between subduction zones will be useful if

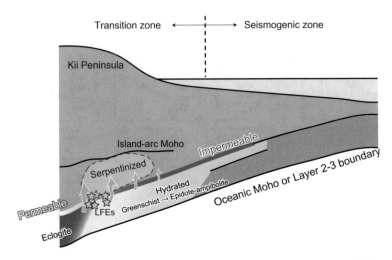

Fig. 6.2 Schematic illustration of the southwestern subduction zone around the Kii Peninsula, according to interpretations made by this study. The sky blue color beneath the plate interface shows the area that we consider to be hydrated, namely the low-velocity zone (LVZ). Differences in LVZ thickness beneath the offshore and onshore regions reflect the differences in the time intervals between the PsL– and PsL+ phases (Fig. 6.1)

we assume that the difference in the fluid process is responsible for the varied features of slow earthquakes. Here, the different fluid processes indicate (i) different amounts of fluid being input to the subduction zones, (ii) different fluid production rates according to the dehydration reaction, (iii) different permeability structures, etc.

Unfortunately, no studies have compared the seismic velocities of LVZs among different subduction zones. Even for the Cascadia and southwestern Japan subduction zones, which are the most well-investigated subduction zones in world, such comparative study has not been performed mainly due to the differences in analysis methods; i.e., tomography analysis has been often applied for southwestern Japan, while RF analyses have been applied for Cascadia (as mentioned earlier, tomography analysis tends to underestimate low-velocity anomalies, so a comparison of velocity structures obtained from tomography and RF analyses is meaningless).

In Sect. 4.4.1, we quantitatively estimated the S-wave velocity of the LVZ beneath the Kii Peninsula to be 2.9–4.2 km/s based on the RF amplitudes of the P-to-S conversion phases. RF inversion analysis as conducted in Chap. 5 has the potential to constrain the LVZ velocity more accurately because the method utilizes information on the differential arrival time between PsL– and PsL+ phases (i.e., not only their amplitudes). Applying a similar method to other subduction zones will help in understanding the different fluid processes among subduction zones.[1]

---

[1] A number of studies have conducted RF analysis in the Cascadia subduction zone (e.g., Audet et al. 2009; Hansen et al. 2012). However, Vp/Vs ratios were estimated instead of S-wave velocity, making it still difficult to compare the results directly with our study. In addition, the estimation

6.3 New Perspectives by High-Frequency Receiver Function Analysis Using OBS Data

In Chap. 5, we demonstrated that high-frequency RF inversion analysis using OBS data detects a thin (<2 km) LVZ along the subducting plate interface and constrains its thickness and S-wave velocity. This success is based on the usage of high-frequency content in the analysis, and special attention to the water and sediment reverberations was necessary to calculate and interpret the high-frequency RFs correctly. It should be noted that this study has, for the first time, clarified S-wave velocity structure near a megathrust fault at a fine scale comparable to active-source seismic surveys. Some previous studies have revealed fine-scale (~0.1–1 km order) P-wave velocity structure along the megathrusts using sophisticated analysis methods (Bangs et al. 2009; Kamei et al. 2012; Li et al. 2015), but this has not been done previously for S-wave velocity structures. Since the Vp/Vs ratio (or Poisson's ratio) provides essential information to quantitatively estimate rock and fluid properties (e.g., Christensen 1984), combining information from active-source seismic surveys and high-frequency RF analysis can help better understand the physical properties of megathrust faults.

Another aspect not considered by this study is anisotropic structure. Anisotropic structure should be dominant near megathrust faults because of margin-normal tectonic loading by a subducting plate (e.g., Lin et al. 2010) and systematic damage brought about by rupture during megathrust earthquakes. Nevertheless, the anisotropic features of megathrust fault have not yet been investigated by active-source surveys; it will require a great number of air-gun shots surrounding a target region, so the experimental cost is very high. RF analysis, in contrast, does not require such expensive experiments because teleseismic ray paths come from various directions. Additive model parameters for determining anisotropic structure make inversion analysis more challenging; however, the usage of transverse-component records (e.g., Frederiksen and Bostock 2000) and more sophisticated inversion algorithms such as trans-dimensional Monte Carlo sampling (e.g., Bodin et al. 2012) can help overcome this problem.

Currently, large-scale scientific projects involving passive seismic observations at offshore sites have been conducted or are underway at subduction zones worldwide (e.g., Scherwath et al. 2011; Kaneda 2014; Toomey et al. 2014). Unfortunately, the application of RF analysis to such offshore data is limited to date (e.g., Audet 2016; Akuhara et al. 2017; Janiszewski and Abers 2015). Our analysis method overcomes the severe problem specific to OBS data by taking into consideration the multiple phases within seawater and sediment layers. We believe that applying a similar analysis to other regions will help us acquire advanced knowledge relating to the physical properties around megathrust faults.

---

bias of seismic velocity resulting from RF analysis is highly dependent on the assumption of the analysis (e.g., dipping layered structure vs. 1-D layered structure), so direct comparison with the results of previous works requires care.

References

Akuhara T, Mochizuki K, Kawakatsu H, Takeuchi N (2017) A fluid-rich layer along the nankai trough megathrust fault off the kii peninsula inferred from receiver function inversion. J Geophys Res Solid Earth 122(8):6524–6537. https://doi.org/10.1002/2017JB013965

Audet P (2016) Receiver functions using obs data: promises and limitations from numerical modelling and examples from the cascadia initiative. Geophys J Int 205(3):1740–1755. https://doi.org/10.1093/gji/ggw111

Audet P, Bostock MG, Christensen NI, Peacock SM (2009) Seismic evidence for overpressured subducted oceanic crust and megathrust fault sealing. Nature 457(7225):76–78. https://doi.org/10.1038/nature07650

Bangs N, Moore G, Gulick S, Pangborn E, Tobin H, Kuramoto S, Taira A (2009) Broad, weak regions of the nankai megathrust and implications for shallow coseismic slip. Earth Planet Sci Lett 284(12):44–49. https://doi.org/10.1016/j.epsl.2009.04.026

Bodin T, Sambridge M, TkaliH, Arroucau P, Gallagher K, Rawlinson N (2012) Transdimensional inversion of receiver functions and surface wave dispersion. J Geophys Res Solid Earth 117(B2):n/a–n/a. https://doi.org/10.1029/2011JB008560

Christensen NI (1984) Pore pressure and oceanic crustal seismic structure. Geophys J Int 79(2):411–423. https://doi.org/10.1111/j.1365-246X.1984.tb02232.x

Frederiksen AW, Bostock MG (2000) Modelling teleseismic waves in dipping anisotropic structures. Geophys J Int 141(2):401–412. https://doi.org/10.1046/j.1365-246x.2000.00090.x

Hansen RT, Bostock MG, Christensen NI (2012) Nature of the low velocity zone in Cascadia from receiver function waveform inversion. Earth Planet Sci Lett 337–338:25–38. https://doi.org/10.1016/j.epsl.2012.05.031

Janiszewski HA, Abers GA (2015) Imaging the plate interface in the cascadia seismogenic zone: new constraints from offshore receiver functions. Seismol Res Lett 86(5):1261–1269. https://doi.org/10.1785/0220150104

Kamei R, Pratt RG, Tsuji T (2012) Waveform tomography imaging of a megasplay fault system in the seismogenic nankai subduction zone. Earth Planet Sci Lett 317(Supplement C):343–353. https://doi.org/10.1016/j.epsl.2011.10.042

Kaneda Y (2014) Donet: a real-time monitoring system for megathrust earthquakes and tsunamis around Southwestern Japan. Oceanography 27

Kato A, Saiga A, Takeda T, Iwasaki T, Matsuzawa T (2014) Non-volcanic seismic swarm and fluid transportation driven by subduction of the Philippine Sea slab beneath the Kii Peninsula, Japan. Earth Planet Space 66(1):86. https://doi.org/10.1186/1880-5981-66-86

Li J, Shillington DJ, Bécel A, Nedimović MR, Webb SC, Saffer DM, Keranen KM, Kuehn H (2015) Downdip variations in seismic reflection character: Implications for fault structure and seismogenic behavior in the Alaska subduction zone. J Geophys Res Solid Earth 120(11):7883–7904. https://doi.org/10.1002/2015JB012338

Lin W, Doan ML, Moore JC, McNeill L, Byrne TB, Ito T, Saffer D, Conin M, Kinoshita M, Sanada Y, Moe KT, Araki E, Tobin H, Boutt D, Kano Y, Hayman NW, Flemings P, Huftile GJ, Cukur D, Buret C, Schleicher AM, Efimenko N, Kawabata K, Buchs DM, Jiang S, Kameo K, Horiguchi K, Wiersberg T, Kopf A, Kitada K, Eguchi N, Toczko S, Takahashi K, Kido Y (2010) Present-day principal horizontal stress orientations in the kumano forearc basin of the southwest japan subduction zone determined from iodp nantroseize drilling site c0009. Geophys Res Lett 37(13):n/a–n/a. https://doi.org/10.1029/2010GL043158

Nedimović MR, Hyndman RD, Ramachandran K, Spence GD (2003) Reflection signature of seismic and aseismic slip on the northern Cascadia subduction interface. Nature 424(6947):416–420. https://doi.org/10.1038/nature01840

Scherwath M, Spence G, Obana K, Kodaira S, Wang K, Riedel M, McGuire J, Collins J (2011) Seafloor seismometers monitor northern cascadia earthquakes. EOS Trans Am Geophys Union 92(47):421–422. https://doi.org/10.1029/2011EO470001

Shiomi K, Matsubara M, Ito Y, Obara K (2008) Simple relationship between seismic activity along Philippine Sea slab and geometry of oceanic Moho beneath southwest Japan. Geophys J Int 173(3):1018–1029. https://doi.org/10.1111/j.1365-246X.2008.03786.x

Toomey D, Allen RM, Barclay AH, Bell SW, Bromirski PD, Carlson RL, Chen X, Collins JA, Dziak RP, Evers B, Forsyth DW, Gerstoft P, Hooft EE, Livelybrooks D, Lodewyk JA, Luther DS, McGuire JJ, Schwartz SY, Tolstoy M, Tréu AM, Weirathmueller M, Wilcock WS (2014) The cascadia initiative: a sea change in seismological studies of subduction zones. Oceanography 27

Chapter 7
Conclusion

In this thesis, we developed a method to calculate receiver functions (RFs) from ocean-bottom seismometer (OBS) data, in which water reverberations on vertical-component records were efficiently removed by applying a frequency domain filter. The filter can be expressed with two parameters: a two-way travel time within the water layer and a reflection coefficient on the seafloor. We presented a nonlinear waveform inversion analysis to determine these parameters from observed data. Then, by applying this method, we investigated the hydrous state of the subducting Philippine Sea Plate around the Kii Peninsula, southwestern Japan.

With the aid of a previous tomography model, we performed time-to-depth conversion of radial RFs to create images of the subsurface structure. From this images, we constructed a 3-D geometry model of the subducting PHS plate using coherent positive and negative RF amplitudes. We confirmed that the negative RF peaks originating from the subducting plate interface extend over the entire study area. We consider that this implies the existence of the LVZ just beneath the plate interface.

RF amplitudes along the plate interface and the oceanic Moho gradually decrease as the slab subducts deeper beneath the Kii Peninsula. This may be explained by the dehydration reactions of the oceanic crust: Eclogitization increases the seismic velocity of the oceanic crust, and accompanying densification causes fractures of the oceanic crust that allow fluid to ascend into the overriding plate. The fluid transferred to the overriding mantle wedge is considered to facilitate serpentinization. An important suggestion here is that the non-volcanic tremors in this region would be characterized by the permeable plate interface. This contrasts with an established idea that long-term slow slip events (SSEs) occur at the impermeable plate interface. We postulate that this difference in the permeability may be a factor that distinguishes the source regions tremors and long-term SSEs.

To further constrain the property of the LVZ along the subducting plate, we performed RF waveform inversion including high-frequency contents for several OBSs. We found that the LVZ is extremely thin (0.2–1.2 km), and its S-wave velocity is quite low (0.7–2.4 km/s). We interpreted this as a fluid-rich layer located at the top

© Springer Nature Singapore Pte Ltd. 2018
T. Akuhara, *Fluid Distribution Along the Nankai-Trough Megathrust Fault off the Kii Peninsula*, Springer Theses, https://doi.org/10.1007/978-981-10-8174-3_7

of the subducting plate. Although our analysis is thus far limited in terms of spatial extent, future investigations into the spatial variations of the LVZ properties will help us better understand fluid distribution along the megathrust fault.

Appendix

Curriculum Vitae

Takeshi Akuhara

Earthquake Research Institute, The University of Tokyo
1-1-1 Yayoi, Bunkyo-ku, Tokyo 113-0032, Japan
e-mail: akuhara@eri.u-tokyo.ac.jp
Web: http://www.eri.u-tokyo.ac.jp/people/akuhara/

Education

- Apr. 2013–Mar. 2016: PhD in Department of Earth Planetary Science, The University of Tokyo, Tokyo, Japan
- Apr. 2011–Mar. 2013: Ms in Department of Earth Planetary Science, The Universiry of Tokyo, Tokyo, Japan

Employment

- Apr. 2017–present: Assistant Professor, Earthquake Research Institute, The University of Tokyo
- Sep. 2016–Mar. 2017: Post Doctoral Fellow, Department of Earth, Ocean and Atmospheric Sciences, The University of British Columbia
- Apr. 2016–Aug. 2016: Post Doctoral Fellow, Earthquake Research Insitute, The University of Tokyo

Major Honors and Awords

- Aug. 2017–Mar. 2019: JSPS Grant-in-Aid for Research Activity start-up
- Sep. 2016–Mar. 2017: JSPS Postdoctral Fellowship for research abroad
- Apr. 2014–Mar. 2016: JSPS Research Fellowship for Young Scientest (for Doctoral Course Students)

© Springer Nature Singapore Pte Ltd. 2018   93
T. Akuhara, *Fluid Distribution Along the Nankai-Trough Megathrust Fault*
off the Kii Peninsula, Springer Theses, https://doi.org/10.1007/978-981-10-8174-3

- Mar. 2016: Reserch Encouragement Award, School of Science, The University of Tokyo
- Oct. 2015: Student Outstanding Presentation Award at Seismological Society of Japan fall meeting
- May. 2015: Student Outstanding Presentation Award at JpGU Meeting 2015

Research Interests

My primary research interest is to understand structural factors that control the slip behavior of megathrust faults in subduction zones, such as megathrust earthquakes and slow earthquakes. For this purpose, I have performed tomography analysis, cluster analysis of earthquakes, and receiver function analysis. My current research topic focuses on (1) developing new methods to estimate physical properties of the megathrust faults from seismic waveform records, (2) understanding fluid processes in subduction zones via comparative studies among subduction zones worldwide, and (3) planning new seismic observations with ocean-bottom seismometers.

Printed in the United States
By Bookmasters